BASIC SKILLS WITH MATH:
A GENERAL REVIEW

A Step-by-Step Approach

JERRY HOWETT

CAMBRIDGE Adult Education
Prentice Hall Regents, Englewood Cliffs, NJ 07632

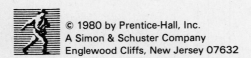

© 1980 by Prentice-Hall, Inc.
A Simon & Schuster Company
Englewood Cliffs, New Jersey 07632

Printed in the United States of America

10 9 8 7

ISBN 0-8428-2119-8

Prentice-Hall International (UK) Limited, *London*
Prentice-Hall of Australia Pty. Limited, *Sydney*
Prentice-Hall Canada Inc., *Toronto*
Prentice-Hall Hispanoamericana, S.A., *Mexico*
Prentice-Hall of India Private Limited, *New Delhi*
Prentice-Hall of Japan, Inc., *Tokyo*
Simon & Schuster Asia Pte. Ltd., *Singapore*
Editora Prentice-Hall do Brasil, Ltda., *Rio de Janeiro*

Contents

Introduction

In this book you will gain skill in working with whole numbers, fractions, decimals, and percents.

Here's how the book will help you. The first step in building math skill is finding out what you need to work on. To help you find out, work the problems called "Step One to Whole Number Skill" starting on page 2. Do all the problems you can. Do not count or use a calculator. Don't worry about the problems that you don't know how to work. This book will help you learn how to work the problems you did not do. Check your answers and fill in the chart on page 6. The right side of the chart lists the pages you need to work on.

Each section of the book begins with problems to help you find the skills you need to practice. There are plenty of practice problems to sharpen your skills. At the end of each section, there are review problems. These will show you how much you have learned.

There is truth in the old saying, "Practice makes perfect." The more you practice, the stronger your skills become. Take your time as you work with this book. The strong skills you build now will help you to solve real-life math problems.

Step One to Whole Number Skill

These problems will help you find out if you need work in the whole number section of this book. Do all the problems you can. Then fill in the chart on page 6 to see which page you should go to next.

In problems 1 to 4 read each number. Then write out the missing word in the name of the number.

1. 7,200   seven _____ , two hundred.

2. 3,080,016   three _____ , eighty _____ , sixteen.

3. 9,504,060   nine _____ , five hundred four _____ , sixty.

4. 120,008,500   one hundred twenty _____ , eight _____ , five hundred.

In problems 5 to 8, read the numbers. Then write the numbers in figures.

5. eight hundred two _____

6. forty thousand, five hundred thirty _____

7. three million, six hundred eight thousand, nine hundred _____

8. sixty million, seventy thousand, fifty _____

9. $\begin{array}{r} 625 \\ + 273 \\ \hline \end{array}$ 10. $\begin{array}{r} 86,147 \\ + 2,532 \\ \hline \end{array}$ 11. $24 + 933 =$ 12. $1,046 + 321 =$

13. $\begin{array}{r} 57 \\ + 89 \\ \hline \end{array}$ 14. $\begin{array}{r} 26 \\ 37 \\ + 94 \\ \hline \end{array}$ 15. $\begin{array}{r} 573 \\ 1,268 \\ 486 \\ + 3,691 \\ \hline \end{array}$ 16. $\begin{array}{r} 9,108 \\ 275 \\ 6,334 \\ + 867 \\ \hline \end{array}$

17. 48 + 276 + 3,197 =   **18.** 6,028 + 55 + 399 =

19. Petra bought paint for $19.95, brushes for $4.29, and paint thinner for $1.85. The sales tax was $1.57. What was her total bill including tax?

20. Al shipped three trunks. The first one weighed 73 pounds. The second weighed 88 pounds. The third weighed 109 pounds. What was the combined weight of the three trunks?

21.
```
  56
- 22
```
22.
```
  897
- 357
```
23.
```
 92,368
-  1,357
```
24.
```
  46
-  8
```

25.
```
  826
- 259
```
26.
```
  62,384
- 29,557
```
27.
```
  804
- 578
```
28.
```
  40,020
- 29,316
```

29. 600 − 46 =   **30.** 2,306 − 728 =

31. 30,000 − 2,907 =   **32.** 10,400 − 6,524 =

33. Juan bought a pair of shoes for $21.65. How much change did he get from $25?

34. The Midvale Adult Center has to sell 2,000 tickets to its fair. 1,437 tickets have been sold so far. How many more tickets have to be sold?

35.
$$\begin{array}{r} 73 \\ \times\ 2 \\ \hline \end{array}$$

36.
$$\begin{array}{r} 423 \\ \times\ 4 \\ \hline \end{array}$$

37.
$$\begin{array}{r} 83 \\ \times\ 21 \\ \hline \end{array}$$

38.
$$\begin{array}{r} 613 \\ \times\ 132 \\ \hline \end{array}$$

39.
$$\begin{array}{r} 96 \\ \times\ 7 \\ \hline \end{array}$$

40.
$$\begin{array}{r} 84 \\ \times\ 56 \\ \hline \end{array}$$

41.
$$\begin{array}{r} 59 \\ \times\ 80 \\ \hline \end{array}$$

42.
$$\begin{array}{r} 427 \\ \times\ 395 \\ \hline \end{array}$$

43. $36 \times 70 =$

44. $489 \times 53 =$

45. $6,024 \times 83 =$

46. $65 \times 2,507 =$

47. $87 \times 10 =$

48. $100 \times 60 =$

49. $48 \times 1000 =$

50. An airplane flies at an average speed of 437 miles per hour. How far can the plane travel in six hours?

51. Sam earns $5 an hour. He works 39 hours a week. How much does he make in a week?

52. $6\overline{)510}$　　　**53.** $9\overline{)666}$　　　**54.** $8\overline{)749}$　　　**55.** $7\overline{)331}$

56. $4\overline{)3,147}$　　**57.** $21\overline{)1,029}$　　**58.** $43\overline{)12,748}$　　**59.** $179\overline{)10,382}$

60. $6,057 \div 9 =$　　　　　　　　**61.** $3,253 \div 8 =$

62. $4,032 \div 72 =$　　　　　　　**63.** $2,433 \div 56 =$

64. Heather can drive an average of 23 miles on a gallon of gas. How many gallons of gas does she need to drive 414 miles?

65. Fred paid $143 for 13 gallons of paint.
What was the price of one gallon?

Check your answers on page 158. Then complete the chart below.

Problem numbers	Number of problems in this section	Number of problems you got right in this section	
1 to 8	8	_____	If you had fewer than 6 problems right, go to page 7.
9 to 20	12	_____	If you had fewer than 9 problems right, go to page 9.
21 to 34	14	_____	If you had fewer than 11 problems right, go to page 18.
35 to 51	17	_____	If you had fewer than 13 problems right, go to page 27.
52 to 65	14	_____	If you had fewer than 11 problems right, go to page 39.

If you missed no more than 13 problems, correct them and go to Step One to Fraction Skill on page 55.

Place Value: Reading and Writing Whole Numbers

Whole numbers are made up of the **digits** 0, 1, 2, 3, 4, 5, 6, 7, 8, and 9. The number 44 has two digits. The number 23,060 has five digits. The value of each digit is different because of its position in the number. Every position has a **place value**. The table below gives the names of the first ten places in our whole number system.

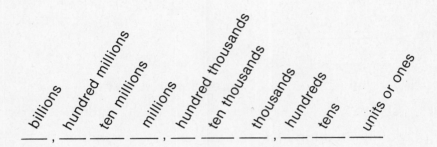

The first 4 in the number 44 (going from left to right) is in the tens place. It has a value of 4 tens or 40. The second 4 is in the units or ones place. It has a value of 4 ones or 4. The digit is still 4, but the value is different because of the digit's place. The 2 in the number 23,060 is in the ten thousands place. It has a value of 2 ten thousands or 20,000. 3 is in the thousands place. It has a value of 3 thousands or 3,000. 6 is in the tens place. It has a value of 6 tens or 60. Notice that the value of the hundreds position and the value of the units position are both 0. The 0 holds the places so the other digits can have the right value.

EXAMPLE: Find the value of 5 in 85,406.

Solution: 5 is in the thousands place. It has a value of 5 thousands or 5,000.

Find the value of each digit shown below. Answers are on page 158.

1. 4 in 2,406 _____

2. 9 in 18,293 _____

3. 7 in 467 _____

4. 3 in 236,025 _____

5. 1 in 321,964 _____

6. 8 in 19,807,253 _____

7. 6 in 36,029,480 _____

8. 2 in 42,980 _____

9. 5 in 499,508 _____

10. 7 in 3,470,000,000 _____

Commas make numbers easier to read. Counting from the right, there is a comma after every three places. Large numbers are read in groups of three. At each comma we say the name of the group of digits that are set off by the comma.

Supply the missing words you need to read each number. Answers are on page 158.

EXAMPLE: 2,043,000   two *million*, forty-three *thousand*.

1. 502   five _____ two.

2. 4,080   four _____ , eighty.

3. 58,320   fifty-eight _____ , three hundred twenty.

4. 6,019,300   six _____ , nineteen_____, three hundred.

5. 246,500   two hundred forty-six _____, five hundred.

6. 3,800   three _____, eight _____.

7. 19,007,200   nineteen _____, seven_____, two _____.

To write whole numbers from words, watch for places that must be held with zeros.

EXAMPLE: Write three million, four hundred eight thousand, six hundred as a whole number.

Solution: This number contains no ten thousands, no tens, and no units. Hold these places with zeros:
3,408,600

Write each of the following as a whole number. Answers are on page 158.

8. three hundred eight _____

9. two hundred sixty-one thousand _____

10. ninety thousand, twenty-four _____

11. four million, one hundred seventy thousand _____

12. eight hundred four thousand, five hundred _____

13. sixty thousand, three hundred _____

14. eleven million, two hundred seven thousand _____

Addition Facts

You can count to get the answers to addition problems. But counting takes too much time. You can work faster if you know the addition facts. Each problem in the exercise below is an addition fact. Do this exercise. Then check your answers. Study the facts you missed until you can do this exercise with no errors. **Answers are on page 158.**

1.
7	9	8	7	9	8	1	5	4	4
+ 3	+ 4	+ 0	+ 9	+ 2	+ 8	+ 5	+ 9	+ 7	+ 1

2.
2	6	7	5	3	9	3	5	2	4
+ 6	+ 3	+ 1	+ 6	+ 1	+ 8	+ 6	+ 5	+ 9	+ 8

3.
5	7	9	3	0	6	2	9	5	1
+ 8	+ 6	+ 0	+ 9	+ 3	+ 4	+ 5	+ 3	+ 2	+ 8

4.
6	3	6	0	1	8	9	4	1	5
+ 5	+ 3	+ 2	+ 2	+ 3	+ 1	+ 6	+ 9	+ 4	+ 3

5.
8	5	8	3	2	6	4	1	6	7
+ 3	+ 1	+ 7	+ 8	+ 7	+ 7	+ 2	+ 2	+ 9	+ 0

6.
7	3	4	1	6	5	2	8	4	5
+ 4	+ 4	+ 0	+ 7	+ 8	+ 7	+ 2	+ 6	+ 5	+ 6

7.
8	6	3	1	8	5	7	2	6	9
+ 5	+ 1	+ 2	+ 1	+ 2	+ 0	+ 8	+ 4	+ 0	+ 5

8.
4	0	5	2	2	9	8	2	6	1
+ 3	+ 1	+ 4	+ 8	+ 3	+ 9	+ 4	+ 1	+ 6	+ 9

9.
7	9	3	4	9	4	3	7	1	7
+ 2	+ 7	+ 5	+ 4	+ 1	+ 6	+ 7	+ 7	+ 6	+ 5

Addition of Larger Numbers

The answer to an addition problem is called the **sum** or **total**. You can find the sum of large numbers with the basic addition facts. Begin by adding all the numbers in the ones column. Continue to add each column until you have finished.

EXAMPLE: Add 46 + 33 =

$$
\begin{array}{r}
46 \\
+\ 33 \\
\hline
79
\end{array}
$$

Step 1.   Add the ones. 6 + 3 = 9.

Step 2.   Add the tens. 4 + 3 = 7.

To check an addition problem, add from the bottom to the top. The new sum should be the same as the old sum.

EXAMPLE: Check the problem above.

Step 1.   Add the ones from the bottom. 3 + 6 = 9.

Step 2.   Add the tens from the bottom. 3 + 4 = 7.

Add and check. Answers are on page 159.

1.
$$\begin{array}{r}27\\+\ 62\\\hline\end{array}\qquad\begin{array}{r}52\\+\ 36\\\hline\end{array}\qquad\begin{array}{r}83\\+\ 13\\\hline\end{array}\qquad\begin{array}{r}13\\+\ 45\\\hline\end{array}\qquad\begin{array}{r}86\\+\ 12\\\hline\end{array}\qquad\begin{array}{r}72\\+\ 26\\\hline\end{array}\qquad\begin{array}{r}34\\+\ 45\\\hline\end{array}$$

2.
$$\begin{array}{r}89\\+\ 10\\\hline\end{array}\qquad\begin{array}{r}43\\+\ 23\\\hline\end{array}\qquad\begin{array}{r}76\\+\ 11\\\hline\end{array}\qquad\begin{array}{r}15\\+\ 74\\\hline\end{array}\qquad\begin{array}{r}42\\+\ 36\\\hline\end{array}\qquad\begin{array}{r}23\\+\ 64\\\hline\end{array}\qquad\begin{array}{r}85\\+\ 14\\\hline\end{array}$$

3.
$$\begin{array}{r}161\\+\ 508\\\hline\end{array}\qquad\begin{array}{r}433\\+\ 360\\\hline\end{array}\qquad\begin{array}{r}757\\+\ 212\\\hline\end{array}\qquad\begin{array}{r}318\\+\ 551\\\hline\end{array}\qquad\begin{array}{r}253\\+\ 704\\\hline\end{array}\qquad\begin{array}{r}280\\+\ 318\\\hline\end{array}$$

4.
$$\begin{array}{r}718\\+\ \ 61\\\hline\end{array}\qquad\begin{array}{r}92\\+\ 603\\\hline\end{array}\qquad\begin{array}{r}333\\+\ \ 54\\\hline\end{array}\qquad\begin{array}{r}38\\+\ 831\\\hline\end{array}\qquad\begin{array}{r}767\\+\ \ 22\\\hline\end{array}\qquad\begin{array}{r}52\\+\ 433\\\hline\end{array}$$

5.
$$\begin{array}{r}6{,}123\\+\ 3{,}542\\\hline\end{array}\qquad\begin{array}{r}4{,}805\\+\ 5{,}073\\\hline\end{array}\qquad\begin{array}{r}8{,}354\\+\ 1{,}123\\\hline\end{array}\qquad\begin{array}{r}6{,}207\\+\ 3{,}281\\\hline\end{array}\qquad\begin{array}{r}4{,}282\\+\ 5{,}017\\\hline\end{array}$$

6.
$$\begin{array}{r}17{,}182\\+\ \ 2{,}615\\\hline\end{array}\qquad\begin{array}{r}81{,}065\\+\ 14{,}302\\\hline\end{array}\qquad\begin{array}{r}54{,}206\\+\ \ 4{,}592\\\hline\end{array}\qquad\begin{array}{r}45{,}192\\+\ 33{,}403\\\hline\end{array}\qquad\begin{array}{r}80{,}279\\+\ \ 9{,}510\\\hline\end{array}$$

If an addition problem is written horizontally (with the numbers standing side by side), rewrite the problem. Put the ones under the ones, the tens under the tens, and so on.

EXAMPLE: Add 23 + 475 =

$$\begin{array}{r} 23 \\ + 475 \\ \hline 498 \end{array}$$

Step 1.   Rewrite the problem.

Step 2.   3 + 5 = 8.

Step 3.   2 + 7 = 9.

Step 4.   nothing + 4 = 4.

Rewrite each problem, add, and check. Answers are on page 159.

7.   24 + 533 =   406 + 52 =   83 + 915 =

8.   614 + 54 =   80 + 617 =   780 + 19 =

9.   124 + 2,325 =   8,607 + 252 =   535 + 6,243 =

10. 5,291 + 407 =   551 + 6,028 =   7,061 + 923 =

11. 91 + 4,905 =   8,543 + 23 =   74 + 2,610 =

12. 1,835 + 41 =   63 + 5,024 =   3,312 + 55 =

13. 157 + 6,540 =   3,892 + 107 =   784 + 7,105 =

Addition with Carrying

When the digits of a column add up to a two-digit number, **carry** the digit left over to the next column to your left.

EXAMPLE: Add 857 + 268 =

$$\begin{array}{r} \overset{1\ 1}{857} \\ +\ \ 268 \\ \hline 1,125 \end{array}$$

Step 1.   Add the ones. 7 + 8 = 15. Write 5 under the ones and carry the 1 to the tens column. The 1 is one ten. It must be added to the tens column.

Step 2.   Add the tens. 1 + 5 = 6. 6 + 6 = 12. Write the 2 under the tens and carry the 1 to the hundreds.

Step 3.   Add the hundreds. 1 + 8 = 9. 9 + 2 = 11.

Add and check. Answers are on page 159.

1.
44	78	37	91	46	73	68
+ 57	+ 24	+ 63	+ 89	+ 98	+ 28	+ 35

2.
15	74	63	84	97	46	67
+ 88	+ 56	+ 79	+ 28	+ 36	+ 99	+ 63

3.
68	87	89	53	62	18	77
+ 46	+ 75	+ 52	+ 27	+ 48	+ 55	+ 66

4.
17	44	38	47	39	24	43
88	60	27	96	53	25	44
+ 53	+ 67	+ 69	+ 80	+ 58	+ 73	+ 19

5.
26	45	59	88	66	56	12
55	32	80	47	25	63	84
+ 66	+ 29	+ 41	+ 70	+ 63	+ 34	+ 57

6.
31	82	46	90	43	57	77
47	68	72	28	87	29	13
+ 58	+ 45	+ 88	+ 16	+ 24	+ 48	+ 46

7.
$$
\begin{array}{r} 341 \\ + 59 \\ \hline \end{array}
\qquad
\begin{array}{r} 228 \\ + 85 \\ \hline \end{array}
\qquad
\begin{array}{r} 368 \\ + 92 \\ \hline \end{array}
\qquad
\begin{array}{r} 625 \\ + 77 \\ \hline \end{array}
\qquad
\begin{array}{r} 439 \\ + 46 \\ \hline \end{array}
\qquad
\begin{array}{r} 773 \\ + 96 \\ \hline \end{array}
$$

8.
$$
\begin{array}{r} 48 \\ + 485 \\ \hline \end{array}
\qquad
\begin{array}{r} 92 \\ + 378 \\ \hline \end{array}
\qquad
\begin{array}{r} 57 \\ + 974 \\ \hline \end{array}
\qquad
\begin{array}{r} 65 \\ + 573 \\ \hline \end{array}
\qquad
\begin{array}{r} 83 \\ + 267 \\ \hline \end{array}
\qquad
\begin{array}{r} 24 \\ + 388 \\ \hline \end{array}
$$

9.
$$
\begin{array}{r} 775 \\ + 638 \\ \hline \end{array}
\qquad
\begin{array}{r} 593 \\ + 549 \\ \hline \end{array}
\qquad
\begin{array}{r} 206 \\ + 297 \\ \hline \end{array}
\qquad
\begin{array}{r} 335 \\ + 866 \\ \hline \end{array}
\qquad
\begin{array}{r} 184 \\ + 499 \\ \hline \end{array}
\qquad
\begin{array}{r} 288 \\ + 827 \\ \hline \end{array}
$$

10.
$$
\begin{array}{r} 725 \\ 87 \\ + 281 \\ \hline \end{array}
\qquad
\begin{array}{r} 968 \\ 75 \\ + 427 \\ \hline \end{array}
\qquad
\begin{array}{r} 458 \\ 56 \\ + 333 \\ \hline \end{array}
\qquad
\begin{array}{r} 166 \\ 43 \\ + 268 \\ \hline \end{array}
\qquad
\begin{array}{r} 925 \\ 79 \\ + 842 \\ \hline \end{array}
\qquad
\begin{array}{r} 943 \\ 58 \\ + 675 \\ \hline \end{array}
$$

11.
$$
\begin{array}{r} 565 \\ 138 \\ + 905 \\ \hline \end{array}
\qquad
\begin{array}{r} 661 \\ 348 \\ + 236 \\ \hline \end{array}
\qquad
\begin{array}{r} 628 \\ 175 \\ + 334 \\ \hline \end{array}
\qquad
\begin{array}{r} 964 \\ 827 \\ + 859 \\ \hline \end{array}
\qquad
\begin{array}{r} 648 \\ 619 \\ + 828 \\ \hline \end{array}
\qquad
\begin{array}{r} 325 \\ 432 \\ + 129 \\ \hline \end{array}
$$

12.
$$
\begin{array}{r} 236 \\ 1{,}940 \\ + 375 \\ \hline \end{array}
\qquad
\begin{array}{r} 808 \\ 2{,}767 \\ + 741 \\ \hline \end{array}
\qquad
\begin{array}{r} 375 \\ 3{,}086 \\ + 829 \\ \hline \end{array}
\qquad
\begin{array}{r} 686 \\ 6{,}421 \\ + 506 \\ \hline \end{array}
\qquad
\begin{array}{r} 327 \\ 8{,}448 \\ + 338 \\ \hline \end{array}
$$

13.
$$
\begin{array}{r} 58 \\ 964 \\ 4{,}277 \\ + 52 \\ \hline \end{array}
\qquad
\begin{array}{r} 23 \\ 158 \\ 3{,}284 \\ + 41 \\ \hline \end{array}
\qquad
\begin{array}{r} 77 \\ 841 \\ 1{,}624 \\ + 12 \\ \hline \end{array}
\qquad
\begin{array}{r} 94 \\ 349 \\ 6{,}953 \\ + 54 \\ \hline \end{array}
\qquad
\begin{array}{r} 79 \\ 926 \\ 1{,}756 \\ + 88 \\ \hline \end{array}
$$

14.
$$
\begin{array}{r} 7{,}222 \\ 843 \\ 8{,}207 \\ + 218 \\ \hline \end{array}
\qquad
\begin{array}{r} 1{,}260 \\ 629 \\ 9{,}324 \\ + 265 \\ \hline \end{array}
\qquad
\begin{array}{r} 2{,}507 \\ 751 \\ 5{,}486 \\ + 770 \\ \hline \end{array}
\qquad
\begin{array}{r} 3{,}445 \\ 544 \\ 2{,}771 \\ + 381 \\ \hline \end{array}
\qquad
\begin{array}{r} 8{,}768 \\ 156 \\ 3{,}258 \\ + 406 \\ \hline \end{array}
$$

Rewrite the following problems with the ones under the ones, the tens under the tens, and so on.

15. 73 + 9 + 494 =   867 + 58 + 4 =

16. 766 + 18 + 604 =   93 + 271 + 48 =

17. 787 + 81 + 5,051 =   46 + 8,432 + 568 =

18. 6,544 + 14 + 2,545 =   1,257 + 672 + 33 =

19. 8 + 60 + 2,592 =   76 + 4,622 + 728 =

20. 5,384 + 6 + 417 =   29 + 566 + 6,788 =

21. 18 + 9,208 + 371 =   5,051 + 39 + 277 =

22. 62 + 583 + 13 + 421 =   780 + 65 + 19 + 271 =

23. 318 + 62 + 872 + 19 =   20 + 47 + 739 + 285 =

24. 39 + 238 + 16 + 2,911 =   318 + 4,207 + 16 + 32 =

25. 4,937 + 12,081 + 736 =   19,763 + 485 + 2,118 =

26. 318 + 9,907 + 24,063 =   7,613 + 24 + 88,552 =

27. 8,016 + 11,238 + 127 =   43 + 1,752 + 18,406 =

28. 79,088 + 314 + 2,607 =   935 + 22,463 + 8,142 =

Addition Applications

To work these problems you must use addition skills. Watch for words like **sum** and **total**. They usually mean to add. Other words such as **combine**, **complete**, **entire**, and **altogether** can sometimes mean to add. Give your answers the correct labels such as $ or miles. Also remember to line up money problems with pennies under pennies, dimes under dimes, and dollars under dollars. **Answers are on page 159.**

EXAMPLE: Find the sum of 29¢, $2.80, and $6.

Solution:
```
  $ .29
    2.80
 +  6.00
  $9.09
```

1. The Midvale Plastics Company employs 673 men and 129 women. What is the total number of employees at the company?

2. This year the Miller family pays $187.50 a month for rent. Their landlord is raising the rent $15 a month for next year. What will be their monthly rent next year?

3. Janet and Tom's phone bill for June listed $8.60 for monthly service, $16.45 for long distance calls, and $.92 for taxes. How much was their total bill?

4. The main floor of the Fox Theater has 516 seats. The balcony has 195 seats. How many seats are there altogether in the theater?

5. In one month the Johnsons paid $150 for rent, $24.63 for the telephone, $18.79 for gas and electricity, and $100 for a loan payment. Find the total of these monthly bills.

6. For lunch Manny had a bowl of chicken soup (207 calories), a ham sandwich (324 calories), coffee with cream (30 calories), and a piece of apple pie (330 calories). What was the total number of calories in Manny's lunch?

7. When Gordon took his car to be fixed at a garage, he had to pay $64.40 for new brakes, $9.39 for a new shock absorber, and $13.66 for an oil change. Find the total for these items.

8. In March Don's Record Shop sold 1,026 record albums. In April they sold 963 albums. In May they sold 1,372 albums. What were the total sales for those three months?

9. In 1940 there were 16,112 people in Midvale. In 1980 there were 28,392 more people in Midvale than in 1940. How many people lived in Midvale in 1980?

10. During a recent election for mayor Mr. Ridley got 9,283 votes. Mr. Green got 4,987 votes. Mr. Munro got 4,062 votes. Find the combined number of votes for these three.

11. Faye's employer makes the following deductions from her weekly paycheck: $25.62 for Social Security, $29.13 for federal income tax, and $14.86 for state income tax. Find the total of these deductions.

12. The distance from New York to Cleveland is 507 miles. The distance from Cleveland to Chicago is 343 miles. What is the distance from New York to Chicago by way of Cleveland?

Subtraction Facts

It takes too much time to count to get the answers to subtraction problems. You can work faster if you know the subtraction facts. Each problem in the exercise below is a subtraction fact. Do this exercise. Check your answers. Then study the facts you missed until you can do this exercise with no errors. **Answers are on page 160.**

1.
$$\begin{array}{r} 8 \\ -2 \end{array} \quad \begin{array}{r} 7 \\ -4 \end{array} \quad \begin{array}{r} 15 \\ -9 \end{array} \quad \begin{array}{r} 14 \\ -6 \end{array} \quad \begin{array}{r} 11 \\ -5 \end{array} \quad \begin{array}{r} 6 \\ -6 \end{array} \quad \begin{array}{r} 12 \\ -3 \end{array} \quad \begin{array}{r} 13 \\ -4 \end{array} \quad \begin{array}{r} 9 \\ -4 \end{array}$$

2.
$$\begin{array}{r} 4 \\ -4 \end{array} \quad \begin{array}{r} 7 \\ -3 \end{array} \quad \begin{array}{r} 11 \\ -9 \end{array} \quad \begin{array}{r} 10 \\ -2 \end{array} \quad \begin{array}{r} 9 \\ -2 \end{array} \quad \begin{array}{r} 10 \\ -9 \end{array} \quad \begin{array}{r} 7 \\ -6 \end{array} \quad \begin{array}{r} 4 \\ -2 \end{array} \quad \begin{array}{r} 5 \\ -4 \end{array}$$

3.
$$\begin{array}{r} 9 \\ -6 \end{array} \quad \begin{array}{r} 3 \\ -2 \end{array} \quad \begin{array}{r} 6 \\ -2 \end{array} \quad \begin{array}{r} 9 \\ -3 \end{array} \quad \begin{array}{r} 12 \\ -9 \end{array} \quad \begin{array}{r} 12 \\ -6 \end{array} \quad \begin{array}{r} 9 \\ -9 \end{array} \quad \begin{array}{r} 10 \\ -7 \end{array} \quad \begin{array}{r} 10 \\ -4 \end{array}$$

4.
$$\begin{array}{r} 6 \\ -4 \end{array} \quad \begin{array}{r} 4 \\ -3 \end{array} \quad \begin{array}{r} 13 \\ -7 \end{array} \quad \begin{array}{r} 3 \\ -1 \end{array} \quad \begin{array}{r} 14 \\ -8 \end{array} \quad \begin{array}{r} 8 \\ -0 \end{array} \quad \begin{array}{r} 7 \\ -7 \end{array} \quad \begin{array}{r} 10 \\ -6 \end{array} \quad \begin{array}{r} 5 \\ -1 \end{array}$$

5.
$$\begin{array}{r} 8 \\ -7 \end{array} \quad \begin{array}{r} 16 \\ -7 \end{array} \quad \begin{array}{r} 8 \\ -4 \end{array} \quad \begin{array}{r} 2 \\ -1 \end{array} \quad \begin{array}{r} 5 \\ -2 \end{array} \quad \begin{array}{r} 14 \\ -5 \end{array} \quad \begin{array}{r} 6 \\ -1 \end{array} \quad \begin{array}{r} 13 \\ -8 \end{array} \quad \begin{array}{r} 1 \\ -1 \end{array}$$

6.
$$\begin{array}{r} 13 \\ -5 \end{array} \quad \begin{array}{r} 12 \\ -8 \end{array} \quad \begin{array}{r} 8 \\ -1 \end{array} \quad \begin{array}{r} 9 \\ -7 \end{array} \quad \begin{array}{r} 15 \\ -6 \end{array} \quad \begin{array}{r} 3 \\ -3 \end{array} \quad \begin{array}{r} 11 \\ -6 \end{array} \quad \begin{array}{r} 11 \\ -3 \end{array} \quad \begin{array}{r} 16 \\ -9 \end{array}$$

7.
$$\begin{array}{r} 18 \\ -9 \end{array} \quad \begin{array}{r} 4 \\ -1 \end{array} \quad \begin{array}{r} 17 \\ -8 \end{array} \quad \begin{array}{r} 8 \\ -6 \end{array} \quad \begin{array}{r} 9 \\ -5 \end{array} \quad \begin{array}{r} 15 \\ -8 \end{array} \quad \begin{array}{r} 7 \\ -2 \end{array} \quad \begin{array}{r} 7 \\ -5 \end{array} \quad \begin{array}{r} 11 \\ -4 \end{array}$$

8.
$$\begin{array}{r} 15 \\ -7 \end{array} \quad \begin{array}{r} 9 \\ -1 \end{array} \quad \begin{array}{r} 7 \\ -1 \end{array} \quad \begin{array}{r} 10 \\ -3 \end{array} \quad \begin{array}{r} 8 \\ -5 \end{array} \quad \begin{array}{r} 13 \\ -6 \end{array} \quad \begin{array}{r} 16 \\ -8 \end{array} \quad \begin{array}{r} 11 \\ -7 \end{array} \quad \begin{array}{r} 2 \\ -2 \end{array}$$

9.
$$\begin{array}{r} 9 \\ -8 \end{array} \quad \begin{array}{r} 14 \\ -7 \end{array} \quad \begin{array}{r} 5 \\ -3 \end{array} \quad \begin{array}{r} 10 \\ -8 \end{array} \quad \begin{array}{r} 8 \\ -3 \end{array} \quad \begin{array}{r} 5 \\ -5 \end{array} \quad \begin{array}{r} 10 \\ -5 \end{array} \quad \begin{array}{r} 12 \\ -5 \end{array} \quad \begin{array}{r} 13 \\ -9 \end{array}$$

10.
$$\begin{array}{r} 17 \\ -9 \end{array} \quad \begin{array}{r} 10 \\ -1 \end{array} \quad \begin{array}{r} 12 \\ -4 \end{array} \quad \begin{array}{r} 6 \\ -3 \end{array} \quad \begin{array}{r} 12 \\ -7 \end{array} \quad \begin{array}{r} 11 \\ -8 \end{array} \quad \begin{array}{r} 14 \\ -9 \end{array} \quad \begin{array}{r} 8 \\ -8 \end{array} \quad \begin{array}{r} 11 \\ -2 \end{array}$$

Subtraction of Larger Numbers

The answer to a subtraction problem is called the **difference.** You can find the difference between two large numbers with the basic subtraction facts. Begin by subtracting the ones. Then subtract the tens. Continue to subtract each column until you have finished.

EXAMPLE: Subtract 86 − 34 =

$$\begin{array}{r} 86 \\ -\ 34 \\ \hline 52 \end{array}$$

Step 1. Subtract the ones. 6 − 4 = 2.

Step 2. Subtract the tens. 8 − 3 = 5.

To check a subtraction problem, add your answer (the difference) to the lower number in the old problem. The sum should equal the top number.

EXAMPLE: Check the problem above.

$$\begin{array}{r} 34 \\ +\ 52 \\ \hline 86 \end{array}$$

Step 1. 4 + 2 = 6.

Step 2. 3 + 5 = 8.

Subtract and check. Answers are on page 160.

1.
$$\begin{array}{r} 87 \\ -\ 25 \end{array}$$
$$\begin{array}{r} 79 \\ -\ 36 \end{array}$$
$$\begin{array}{r} 36 \\ -\ 24 \end{array}$$
$$\begin{array}{r} 65 \\ -\ 40 \end{array}$$
$$\begin{array}{r} 56 \\ -\ 26 \end{array}$$
$$\begin{array}{r} 89 \\ -\ 33 \end{array}$$
$$\begin{array}{r} 75 \\ -\ 14 \end{array}$$

2.
$$\begin{array}{r} 883 \\ -\ 442 \end{array}$$
$$\begin{array}{r} 736 \\ -\ 603 \end{array}$$
$$\begin{array}{r} 956 \\ -\ 451 \end{array}$$
$$\begin{array}{r} 779 \\ -\ 238 \end{array}$$
$$\begin{array}{r} 265 \\ -\ 162 \end{array}$$
$$\begin{array}{r} 544 \\ -\ 314 \end{array}$$

3.
$$\begin{array}{r} 855 \\ -\ 214 \end{array}$$
$$\begin{array}{r} 726 \\ -\ 322 \end{array}$$
$$\begin{array}{r} 843 \\ -\ 621 \end{array}$$
$$\begin{array}{r} 672 \\ -\ 440 \end{array}$$
$$\begin{array}{r} 796 \\ -\ 275 \end{array}$$
$$\begin{array}{r} 548 \\ -\ 328 \end{array}$$

4.
$$\begin{array}{r} 13{,}082 \\ -\ 2{,}061 \end{array}$$
$$\begin{array}{r} 75{,}164 \\ -\ 4{,}051 \end{array}$$
$$\begin{array}{r} 32{,}589 \\ -\ 1{,}445 \end{array}$$
$$\begin{array}{r} 76{,}625 \\ -\ 3{,}515 \end{array}$$
$$\begin{array}{r} 91{,}743 \\ -\ 1{,}241 \end{array}$$

5.
$$\begin{array}{r} 77{,}254 \\ -\ 62{,}033 \end{array}$$
$$\begin{array}{r} 67{,}381 \\ -\ 14{,}240 \end{array}$$
$$\begin{array}{r} 36{,}893 \\ -\ 13{,}522 \end{array}$$
$$\begin{array}{r} 49{,}842 \\ -\ 26{,}301 \end{array}$$
$$\begin{array}{r} 23{,}558 \\ -\ 11{,}417 \end{array}$$

Subtraction with Borrowing

When the bottom number in any column is too large to subtract from the top number, **borrow** from the next column in the top number.

EXAMPLE: Subtract 85 − 39 =

$$\begin{array}{r} {\scriptstyle 7\ 15} \\ \cancel{85} \\ -\ 39 \\ \hline 46 \end{array}$$

Step 1.   9 is too large to subtract from 5. Borrow 1 ten from the tens column (8 − 1 = 7) and add it to the units (10 + 5 = 15). Cross out the 8 and write 7 above it. Cross out the 5 and write 15 above it.

Step 2.   Subtract the units. 15 − 9 = 6.

Step 3.   Subtract the tens. 7 − 3 = 4.

Step 4.   Check. 39 + 46 = 85.

Subtract and check. Answers are on page 160.

1.

33	96	24	78	86	61	45
− 7	− 9	− 6	− 9	− 8	− 4	− 7

2.

83	43	66	42	25	77	36
− 8	− 9	− 7	− 9	− 6	− 8	− 8

3.

58	96	47	58	51	32	93
− 29	− 48	− 28	− 19	− 43	− 16	− 65

4.

52	64	25	87	91	33	46
− 25	− 38	− 16	− 18	− 55	− 16	− 28

5.

86	67	30	42	71	52	93
− 18	− 29	− 16	− 15	− 47	− 23	− 18

In some problems you will need to borrow several times. Study the following example carefully.

EXAMPLE:
$$\begin{array}{r} \overset{6\ 14\ 27\ \overset{17}{\ }16}{7\,4,3\,8\,6} \\ -\ 2\,5,1\,9\,7 \\ \hline 4\,9,1\,8\,9 \end{array}$$

Check:
$$\begin{array}{r} 25,197 \\ +\ 49,189 \\ \hline 74,386 \end{array}$$

Subtract and check. Answers are on page 160.

6.
$$\begin{array}{r} 564 \\ -\ 85 \end{array} \qquad \begin{array}{r} 414 \\ -\ 56 \end{array} \qquad \begin{array}{r} 666 \\ -\ 79 \end{array} \qquad \begin{array}{r} 587 \\ -\ 88 \end{array} \qquad \begin{array}{r} 243 \\ -\ 94 \end{array} \qquad \begin{array}{r} 371 \\ -\ 75 \end{array}$$

7.
$$\begin{array}{r} 811 \\ -\ 243 \end{array} \qquad \begin{array}{r} 572 \\ -\ 418 \end{array} \qquad \begin{array}{r} 467 \\ -\ 199 \end{array} \qquad \begin{array}{r} 340 \\ -\ 238 \end{array} \qquad \begin{array}{r} 551 \\ -\ 365 \end{array} \qquad \begin{array}{r} 760 \\ -\ 467 \end{array}$$

8.
$$\begin{array}{r} 6,175 \\ -\ 496 \end{array} \qquad \begin{array}{r} 7,240 \\ -\ 384 \end{array} \qquad \begin{array}{r} 5,628 \\ -\ 979 \end{array} \qquad \begin{array}{r} 6,425 \\ -\ 556 \end{array} \qquad \begin{array}{r} 3,183 \\ -\ 287 \end{array}$$

9.
$$\begin{array}{r} 4,236 \\ -\ 1,448 \end{array} \qquad \begin{array}{r} 5,668 \\ -\ 2,699 \end{array} \qquad \begin{array}{r} 6,673 \\ -\ 3,887 \end{array} \qquad \begin{array}{r} 4,290 \\ -\ 2,947 \end{array} \qquad \begin{array}{r} 3,837 \\ -\ 2,608 \end{array}$$

10.
$$\begin{array}{r} 5,335 \\ -\ 2,914 \end{array} \qquad \begin{array}{r} 2,414 \\ -\ 1,671 \end{array} \qquad \begin{array}{r} 7,380 \\ -\ 4,093 \end{array} \qquad \begin{array}{r} 9,127 \\ -\ 3,882 \end{array} \qquad \begin{array}{r} 8,358 \\ -\ 6,169 \end{array}$$

11.
$$\begin{array}{r} 15,624 \\ -\ 9,587 \end{array} \qquad \begin{array}{r} 34,124 \\ -\ 5,025 \end{array} \qquad \begin{array}{r} 86,472 \\ -\ 9,913 \end{array} \qquad \begin{array}{r} 27,630 \\ -\ 6,472 \end{array} \qquad \begin{array}{r} 81,862 \\ -\ 3,389 \end{array}$$

12.
$$\begin{array}{r} 24,125 \\ -\ 14,666 \end{array} \qquad \begin{array}{r} 58,168 \\ -\ 39,202 \end{array} \qquad \begin{array}{r} 61,246 \\ -\ 44,425 \end{array} \qquad \begin{array}{r} 36,236 \\ -\ 21,379 \end{array} \qquad \begin{array}{r} 45,611 \\ -\ 16,920 \end{array}$$

Borrowing the Zeros

Borrowing with zeros in the top number looks hard, but it isn't. Just borrow from the first digit in the top number that is not a zero.

EXAMPLE: Subtract 904 − 356 =

$\overset{8\ 10}{\cancel{90}4}$
$-\ 356$

Step 1.   You cannot subtract 6 from 4. Borrow 1 hundred from the 9 in the hundreds place (9 − 1 = 8). You now have 10 tens in the tens place and 8 hundreds in the hundreds place.

$\overset{\ \ 9}{\overset{8\ \cancel{10}\ 14}{\cancel{904}}}$
$-\ 356$
$\overline{\ 548}$

Step 2.   Borrow 1 ten from the 10 in the tens place (10 − 1 = 9). You now have 14 units in the units place and 9 tens in the tens places.

Step 3.   Subtract the units. 14 − 6 = 8.

Step 4.   Subtract the tens. 9 − 5 = 4.

Step 5.   Subtract the hundreds. 8 − 3 = 5.

Step 6.   Check. 356 + 548 = 904.

Subtract and check. Answers are on page 160.

1.	602	904	305	208	407	606
	− 58	− 27	− 66	− 49	− 58	− 29

2.	801	407	503	808	506	707
	− 236	− 209	− 388	− 619	− 218	− 379

3.	402	308	106	403	108	303
	− 55	− 69	− 77	− 45	− 79	− 46

4.	706	503	802	901	704	206
	− 267	− 125	− 708	− 355	− 477	− 168

In some problems you will need to borrow from more than one zero. Study the following example carefully.

EXAMPLE: Subtract 9,004 − 3,125 =

$$\begin{array}{r} \overset{\scriptstyle 8}{\cancel{9}},\overset{\scriptstyle 10}{\cancel{0}}04 \\ -\ 3,125 \end{array}$$

Step 1. You cannot subtract 5 from 4. Borrow 1 thousand from the 9 in the thousands place (9 − 1 = 8). You now have 10 hundreds in the hundreds place.

$$\begin{array}{r} \overset{\scriptstyle 8}{\cancel{9}},\overset{\scriptstyle 9}{\overset{\scriptstyle 10}{\cancel{0}}}\overset{\scriptstyle 10}{\cancel{0}}4 \\ -\ 3,125 \end{array}$$

Step 2. Borrow 1 hundred from the 10 in the hundreds place (10 − 1 = 9). You now have 10 tens in the tens place.

Step 3. Borrow 1 ten from the 10 in the tens place (10 − 1 = 9). You now have 14 units in the units place.

$$\begin{array}{r} \overset{\scriptstyle 8}{\cancel{9}},\overset{\scriptstyle 9}{\cancel{0}}\overset{\scriptstyle 9}{\cancel{0}}\overset{\scriptstyle 14}{\cancel{4}} \\ -\ 3,125 \\ \hline 5,879 \end{array}$$

Step 4. Subtract the units. 14 − 5 = 9.

Step 5. Subtract the tens. 9 − 2 = 7.

Step 6. Subtract the hundreds. 9 − 1 = 8.

Step 7. Subtract the thousands. 8 − 3 = 5.

Step 8. Check. 3,125 + 5,879 = 9,004.

Subtract and check. Answers are on page 160.

1.
$$\begin{array}{r} 600 \\ -\ 226 \end{array} \qquad \begin{array}{r} 800 \\ -\ 354 \end{array} \qquad \begin{array}{r} 500 \\ -\ 189 \end{array} \qquad \begin{array}{r} 900 \\ -\ 614 \end{array} \qquad \begin{array}{r} 300 \\ -\ 108 \end{array} \qquad \begin{array}{r} 400 \\ -\ 376 \end{array}$$

2.
$$\begin{array}{r} 4,030 \\ -\ 2,945 \end{array} \qquad \begin{array}{r} 2,020 \\ -\ 1,167 \end{array} \qquad \begin{array}{r} 7,060 \\ -\ 1,243 \end{array} \qquad \begin{array}{r} 5,070 \\ -\ 3,467 \end{array} \qquad \begin{array}{r} 3,080 \\ -\ 1,096 \end{array} \qquad \begin{array}{r} 4,060 \\ -\ 2,388 \end{array}$$

3.
$$\begin{array}{r} 4,000 \\ -\ 1,256 \end{array} \qquad \begin{array}{r} 3,000 \\ -\ 2,338 \end{array} \qquad \begin{array}{r} 8,000 \\ -\ 4,411 \end{array} \qquad \begin{array}{r} 2,000 \\ -\ 1,950 \end{array} \qquad \begin{array}{r} 7,000 \\ -\ 3,076 \end{array} \qquad \begin{array}{r} 6,000 \\ -\ 2,447 \end{array}$$

4.
$$\begin{array}{r} 80,500 \\ -\ 23,475 \end{array} \qquad \begin{array}{r} 60,070 \\ -\ 29,083 \end{array} \qquad \begin{array}{r} 40,009 \\ -\ 16,668 \end{array} \qquad \begin{array}{r} 23,000 \\ -\ 18,367 \end{array} \qquad \begin{array}{r} 90,060 \\ -\ 24,077 \end{array}$$

Horizontal Subtraction

Rewrite the following problems with units under units, tens under tens, and so on.

Subtract and check. Answers are on page 160.

1. $800 - 57 =$ $300 - 126 =$ $700 - 49 =$

2. $423 - 297 =$ $854 - 368 =$ $431 - 178 =$

3. $7,000 - 1,270 =$ $3,000 - 1,681 =$ $9,000 - 4,023 =$

4. $9,050 - 387 =$ $6,040 - 833 =$ $5,080 - 782 =$

5. $3,003 - 558 =$ $2,007 - 399 =$ $4,008 - 470 =$

6. $19,236 - 6,059 =$ $73,198 - 2,909 =$ $67,218 - 4,979 =$

7. $18,007 - 5,668 =$ $20,050 - 9,266 =$ $30,600 - 9,482 =$

8. $90,040 - 18,255 =$ $60,000 - 13,478 =$ $40,005 - 20,386 =$

Subtraction Applications

To work these problems, you must use subtraction skills. Watch for words like **difference** and **balance**. They usually mean to subtract. Word groups such as **how much greater** and **how much less** also usually mean to subtract. Give your answers the correct labels such as $ or pounds. Remember to line up money problems with pennies under pennies, dimes under dimes, and dollars under dollars. **Answers are on page 161.**

1. Tom makes $235 a week. His employer deducts $37.85 for taxes and Social Security. How much does Tom take home each week?

2. The Hicks family bought a house for $32,500. They made a down payment of $4,780. How much more do they owe for the house?

3. 683 people signed up to go on a trip to Miami. 506 people actually went on the trip. How many people who signed up did not go?

4. Mr. and Mrs. Meyer must drive 413 miles to get to their son's house. They stopped to eat after they had driven 225 miles. How much farther do they have to drive?

5. Arkansas became a state in 1836. For how many years had Arkansas been a state in 1980?

6. The town of Midvale wants to raise $850,000 to build a new health center. They have collected $473,260. How much more money do they need?

7. Dan bought a pair of jeans for $16.89. How much change did he get from $20?

8. Frank borrowed $2,800 to buy a used car. So far he has paid back $1,675. How much more does Frank owe on the loan?

9. Sam Brown earns $14,800 a year. Jane Brown earns $10,980 a year. Find the difference between their salaries.

10. The Globe Theater holds 420 people. At a Saturday night show 87 seats were empty. How many people were at the show that night?

11. A tape deck once sold for $129. It is on sale for $89.85. How much can you save by buying the tape deck on sale?

12. In 1970 there were 129,368 people in Central County. In 1980 there were 201,245 people. How many more people lived in Central County in 1980?

13. In a day Kuwait produces about 2,150,000 barrels of oil. Saudi Arabia produces about 9,500,000 barrels a day. In a day how much more oil does Saudi Arabia produce than Kuwait?

14. The Garcias bought new living room furniture for $480. They got a discount of $35 for paying cash. What was the price of the furniture after the discount?

Multiplication Facts

You could add to get multiplication answers, but it would be very slow. Multiplication is a fast method of adding. To multiply well, you must know the multiplication facts. Each problem in the exercise below is a multiplication fact. Go through this exercise without adding. Check your answers on page 161. Then study the facts you missed until you can do this exercise with no errors.

1. $\begin{array}{r}3\\\times 7\end{array}$ $\begin{array}{r}6\\\times 2\end{array}$ $\begin{array}{r}11\\\times 7\end{array}$ $\begin{array}{r}3\\\times 2\end{array}$ $\begin{array}{r}4\\\times 12\end{array}$ $\begin{array}{r}7\\\times 0\end{array}$ $\begin{array}{r}6\\\times 9\end{array}$ $\begin{array}{r}1\\\times 2\end{array}$

2. $\begin{array}{r}6\\\times 6\end{array}$ $\begin{array}{r}9\\\times 12\end{array}$ $\begin{array}{r}6\\\times 0\end{array}$ $\begin{array}{r}11\\\times 5\end{array}$ $\begin{array}{r}12\\\times 10\end{array}$ $\begin{array}{r}3\\\times 12\end{array}$ $\begin{array}{r}4\\\times 1\end{array}$ $\begin{array}{r}4\\\times 7\end{array}$

3. $\begin{array}{r}2\\\times 8\end{array}$ $\begin{array}{r}7\\\times 6\end{array}$ $\begin{array}{r}10\\\times 8\end{array}$ $\begin{array}{r}9\\\times 2\end{array}$ $\begin{array}{r}11\\\times 3\end{array}$ $\begin{array}{r}8\\\times 4\end{array}$ $\begin{array}{r}1\\\times 7\end{array}$ $\begin{array}{r}5\\\times 11\end{array}$

4. $\begin{array}{r}11\\\times 8\end{array}$ $\begin{array}{r}9\\\times 0\end{array}$ $\begin{array}{r}2\\\times 12\end{array}$ $\begin{array}{r}7\\\times 2\end{array}$ $\begin{array}{r}5\\\times 7\end{array}$ $\begin{array}{r}12\\\times 7\end{array}$ $\begin{array}{r}9\\\times 3\end{array}$ $\begin{array}{r}4\\\times 9\end{array}$

5. $\begin{array}{r}6\\\times 12\end{array}$ $\begin{array}{r}9\\\times 9\end{array}$ $\begin{array}{r}8\\\times 6\end{array}$ $\begin{array}{r}0\\\times 5\end{array}$ $\begin{array}{r}8\\\times 11\end{array}$ $\begin{array}{r}4\\\times 2\end{array}$ $\begin{array}{r}5\\\times 2\end{array}$ $\begin{array}{r}7\\\times 9\end{array}$

6. $\begin{array}{r}5\\\times 12\end{array}$ $\begin{array}{r}8\\\times 2\end{array}$ $\begin{array}{r}10\\\times 1\end{array}$ $\begin{array}{r}2\\\times 11\end{array}$ $\begin{array}{r}10\\\times 3\end{array}$ $\begin{array}{r}11\\\times 10\end{array}$ $\begin{array}{r}8\\\times 12\end{array}$ $\begin{array}{r}6\\\times 8\end{array}$

7. $\begin{array}{r}5\\\times 4\end{array}$ $\begin{array}{r}3\\\times 5\end{array}$ $\begin{array}{r}10\\\times 6\end{array}$ $\begin{array}{r}3\\\times 3\end{array}$ $\begin{array}{r}1\\\times 9\end{array}$ $\begin{array}{r}2\\\times 2\end{array}$ $\begin{array}{r}6\\\times 3\end{array}$ $\begin{array}{r}4\\\times 11\end{array}$

8. $\begin{array}{r}12\\\times 8\end{array}$ $\begin{array}{r}10\\\times 11\end{array}$ $\begin{array}{r}9\\\times 8\end{array}$ $\begin{array}{r}8\\\times 8\end{array}$ $\begin{array}{r}7\\\times 10\end{array}$ $\begin{array}{r}4\\\times 4\end{array}$ $\begin{array}{r}4\\\times 6\end{array}$ $\begin{array}{r}7\\\times 3\end{array}$

9.
$$\begin{array}{r} 11 \\ \times 11 \\ \hline \end{array} \quad \begin{array}{r} 10 \\ \times 4 \\ \hline \end{array} \quad \begin{array}{r} 6 \\ \times 5 \\ \hline \end{array} \quad \begin{array}{r} 4 \\ \times 8 \\ \hline \end{array} \quad \begin{array}{r} 0 \\ \times 11 \\ \hline \end{array} \quad \begin{array}{r} 9 \\ \times 5 \\ \hline \end{array} \quad \begin{array}{r} 5 \\ \times 8 \\ \hline \end{array} \quad \begin{array}{r} 12 \\ \times 3 \\ \hline \end{array}$$

10.
$$\begin{array}{r} 0 \\ \times 8 \\ \hline \end{array} \quad \begin{array}{r} 12 \\ \times 6 \\ \hline \end{array} \quad \begin{array}{r} 11 \\ \times 9 \\ \hline \end{array} \quad \begin{array}{r} 8 \\ \times 3 \\ \hline \end{array} \quad \begin{array}{r} 7 \\ \times 12 \\ \hline \end{array} \quad \begin{array}{r} 3 \\ \times 11 \\ \hline \end{array} \quad \begin{array}{r} 9 \\ \times 6 \\ \hline \end{array} \quad \begin{array}{r} 6 \\ \times 4 \\ \hline \end{array}$$

11.
$$\begin{array}{r} 6 \\ \times 10 \\ \hline \end{array} \quad \begin{array}{r} 7 \\ \times 4 \\ \hline \end{array} \quad \begin{array}{r} 1 \\ \times 4 \\ \hline \end{array} \quad \begin{array}{r} 10 \\ \times 9 \\ \hline \end{array} \quad \begin{array}{r} 11 \\ \times 12 \\ \hline \end{array} \quad \begin{array}{r} 4 \\ \times 5 \\ \hline \end{array} \quad \begin{array}{r} 11 \\ \times 0 \\ \hline \end{array} \quad \begin{array}{r} 3 \\ \times 6 \\ \hline \end{array}$$

12.
$$\begin{array}{r} 3 \\ \times 9 \\ \hline \end{array} \quad \begin{array}{r} 10 \\ \times 5 \\ \hline \end{array} \quad \begin{array}{r} 3 \\ \times 4 \\ \hline \end{array} \quad \begin{array}{r} 6 \\ \times 11 \\ \hline \end{array} \quad \begin{array}{r} 5 \\ \times 1 \\ \hline \end{array} \quad \begin{array}{r} 2 \\ \times 1 \\ \hline \end{array} \quad \begin{array}{r} 10 \\ \times 2 \\ \hline \end{array} \quad \begin{array}{r} 2 \\ \times 4 \\ \hline \end{array}$$

13.
$$\begin{array}{r} 12 \\ \times 4 \\ \hline \end{array} \quad \begin{array}{r} 4 \\ \times 10 \\ \hline \end{array} \quad \begin{array}{r} 8 \\ \times 10 \\ \hline \end{array} \quad \begin{array}{r} 11 \\ \times 4 \\ \hline \end{array} \quad \begin{array}{r} 5 \\ \times 5 \\ \hline \end{array} \quad \begin{array}{r} 5 \\ \times 10 \\ \hline \end{array} \quad \begin{array}{r} 4 \\ \times 3 \\ \hline \end{array} \quad \begin{array}{r} 6 \\ \times 7 \\ \hline \end{array}$$

14.
$$\begin{array}{r} 7 \\ \times 7 \\ \hline \end{array} \quad \begin{array}{r} 7 \\ \times 11 \\ \hline \end{array} \quad \begin{array}{r} 2 \\ \times 7 \\ \hline \end{array} \quad \begin{array}{r} 3 \\ \times 10 \\ \hline \end{array} \quad \begin{array}{r} 0 \\ \times 12 \\ \hline \end{array} \quad \begin{array}{r} 1 \\ \times 3 \\ \hline \end{array} \quad \begin{array}{r} 11 \\ \times 2 \\ \hline \end{array} \quad \begin{array}{r} 10 \\ \times 12 \\ \hline \end{array}$$

15.
$$\begin{array}{r} 1 \\ \times 10 \\ \hline \end{array} \quad \begin{array}{r} 8 \\ \times 0 \\ \hline \end{array} \quad \begin{array}{r} 2 \\ \times 6 \\ \hline \end{array} \quad \begin{array}{r} 1 \\ \times 1 \\ \hline \end{array} \quad \begin{array}{r} 2 \\ \times 3 \\ \hline \end{array} \quad \begin{array}{r} 9 \\ \times 11 \\ \hline \end{array} \quad \begin{array}{r} 10 \\ \times 7 \\ \hline \end{array} \quad \begin{array}{r} 8 \\ \times 9 \\ \hline \end{array}$$

16.
$$\begin{array}{r} 10 \\ \times 10 \\ \hline \end{array} \quad \begin{array}{r} 12 \\ \times 1 \\ \hline \end{array} \quad \begin{array}{r} 7 \\ \times 5 \\ \hline \end{array} \quad \begin{array}{r} 7 \\ \times 8 \\ \hline \end{array} \quad \begin{array}{r} 11 \\ \times 6 \\ \hline \end{array} \quad \begin{array}{r} 9 \\ \times 7 \\ \hline \end{array} \quad \begin{array}{r} 2 \\ \times 9 \\ \hline \end{array} \quad \begin{array}{r} 5 \\ \times 9 \\ \hline \end{array}$$

17.
$$\begin{array}{r} 9 \\ \times 4 \\ \hline \end{array} \quad \begin{array}{r} 8 \\ \times 7 \\ \hline \end{array} \quad \begin{array}{r} 3 \\ \times 0 \\ \hline \end{array} \quad \begin{array}{r} 12 \\ \times 9 \\ \hline \end{array} \quad \begin{array}{r} 12 \\ \times 5 \\ \hline \end{array} \quad \begin{array}{r} 9 \\ \times 10 \\ \hline \end{array} \quad \begin{array}{r} 12 \\ \times 12 \\ \hline \end{array} \quad \begin{array}{r} 5 \\ \times 3 \\ \hline \end{array}$$

18.
$$\begin{array}{r} 12 \\ \times 4 \\ \hline \end{array} \quad \begin{array}{r} 12 \\ \times 11 \\ \hline \end{array} \quad \begin{array}{r} 2 \\ \times 5 \\ \hline \end{array} \quad \begin{array}{r} 3 \\ \times 8 \\ \hline \end{array} \quad \begin{array}{r} 2 \\ \times 10 \\ \hline \end{array} \quad \begin{array}{r} 5 \\ \times 6 \\ \hline \end{array} \quad \begin{array}{r} 8 \\ \times 5 \\ \hline \end{array} \quad \begin{array}{r} 1 \\ \times 6 \\ \hline \end{array}$$

The Multiplication Table

The facts in the multiplication table are among the most important building blocks of mathematics. On this page is the multiplication table from 1 to 12. To find a multiplication fact, pick one number in the left-hand column. Then find the other number in the top row. From the left-hand number, follow across the row. From the top number, follow down the column. Where the row and the column meet is the answer. For example, to find 7 times 9, find 7 at the left and 9 at the top of the table. Follow 7 across and 9 down until they meet. They meet at 63. $7 \times 9 = 63$.

The table is meant to be used as a study tool not as a crutch. Do not depend on this table to do problems in this book. Take the time to memorize the facts in the table. Do not try to memorize the entire table all at once. Memorize one column at a time. If you do, you will find the rest of the material in this book much easier.

	1	2	3	4	5	6	7	8	9	10	11	12
1	1	2	3	4	5	6	7	8	9	10	11	12
2	2	4	6	8	10	12	14	16	18	20	22	24
3	3	6	9	12	15	18	21	24	27	30	33	36
4	4	8	12	16	20	24	28	32	36	40	44	48
5	5	10	15	20	25	30	35	40	45	50	55	60
6	6	12	18	24	30	36	42	48	54	60	66	72
7	7	14	21	28	35	42	49	56	63	70	77	84
8	8	16	24	32	40	48	56	64	72	80	88	96
9	9	18	27	36	45	54	63	72	81	90	99	108
10	10	20	30	40	50	60	70	80	90	100	110	120
11	11	22	33	44	55	66	77	88	99	110	121	132
12	12	24	36	48	60	72	84	96	108	120	132	144

Multiplication of Larger Numbers

The answer to a multiplication problem is called the **product.** To find the product of a large number and a one-digit number begin at the right. Multiply every digit of the large number by the one-digit number.

EXAMPLE: Multiply 732 × 3 =

$$\begin{array}{r} 732 \\ \times\ \ 3 \\ \hline 2,196 \end{array}$$

Step 1.   3 × 2 = 6.

Step 2.   3 × 3 = 9.

Step 3.   3 × 7 = 21.

There is no easy way to check a multiplication problem. You can divide your answer by the bottom number in the problem. If you get the top number, your answer is correct. An easier method is to repeat the steps and watch for mistakes. It is a good idea in all math problems to check every step you have finished.

Multiply and check. Answers are on page 161.

1.
$$\begin{array}{r} 82 \\ \times\ 4 \\ \hline \end{array}$$
$$\begin{array}{r} 63 \\ \times\ 3 \\ \hline \end{array}$$
$$\begin{array}{r} 74 \\ \times\ 2 \\ \hline \end{array}$$
$$\begin{array}{r} 90 \\ \times\ 9 \\ \hline \end{array}$$
$$\begin{array}{r} 52 \\ \times\ 4 \\ \hline \end{array}$$
$$\begin{array}{r} 71 \\ \times\ 8 \\ \hline \end{array}$$
$$\begin{array}{r} 84 \\ \times\ 2 \\ \hline \end{array}$$

2.
$$\begin{array}{r} 41 \\ \times\ 7 \\ \hline \end{array}$$
$$\begin{array}{r} 30 \\ \times\ 8 \\ \hline \end{array}$$
$$\begin{array}{r} 55 \\ \times\ 1 \\ \hline \end{array}$$
$$\begin{array}{r} 62 \\ \times\ 4 \\ \hline \end{array}$$
$$\begin{array}{r} 43 \\ \times\ 3 \\ \hline \end{array}$$
$$\begin{array}{r} 72 \\ \times\ 4 \\ \hline \end{array}$$
$$\begin{array}{r} 70 \\ \times\ 9 \\ \hline \end{array}$$

3.
$$\begin{array}{r} 622 \\ \times\ 4 \\ \hline \end{array}$$
$$\begin{array}{r} 301 \\ \times\ 8 \\ \hline \end{array}$$
$$\begin{array}{r} 432 \\ \times\ 3 \\ \hline \end{array}$$
$$\begin{array}{r} 324 \\ \times\ 2 \\ \hline \end{array}$$
$$\begin{array}{r} 811 \\ \times\ 9 \\ \hline \end{array}$$
$$\begin{array}{r} 923 \\ \times\ 1 \\ \hline \end{array}$$

4.
$$\begin{array}{r} 308 \\ \times\ 1 \\ \hline \end{array}$$
$$\begin{array}{r} 904 \\ \times\ 2 \\ \hline \end{array}$$
$$\begin{array}{r} 730 \\ \times\ 2 \\ \hline \end{array}$$
$$\begin{array}{r} 422 \\ \times\ 4 \\ \hline \end{array}$$
$$\begin{array}{r} 611 \\ \times\ 8 \\ \hline \end{array}$$
$$\begin{array}{r} 720 \\ \times\ 4 \\ \hline \end{array}$$

5.
$$\begin{array}{r} 1,233 \\ \times\ 2 \\ \hline \end{array}$$
$$\begin{array}{r} 4,011 \\ \times\ 8 \\ \hline \end{array}$$
$$\begin{array}{r} 9,111 \\ \times\ 7 \\ \hline \end{array}$$
$$\begin{array}{r} 3,234 \\ \times\ 2 \\ \hline \end{array}$$
$$\begin{array}{r} 6,033 \\ \times\ 3 \\ \hline \end{array}$$

Multiplication by Two- and Three-Digit Numbers

To find the product of two large numbers, also begin at the right. Multiply each digit of the top number by the units of the bottom number. The first digit of this **partial product** should be directly under the units column. Then multiply each digit of the top number by the tens digit in the bottom number. The first digit of this partial product should be directly under the tens column. Continue until you have a partial product for each digit in the bottom number. Add the partial products to get the final product.

EXAMPLE: Multiply 422 × 231 =

```
      422
  ×   231
      422
   12 66
   84 4
   97,482
```

Step 1.   The first partial product is 1 × 422 = 422.

Step 2.   The second partial product is 3 × 422 = 1266.

Step 3.   The third partial product is 2 × 422 = 844. Notice that the first 4 of 844 is under the hundreds column.

Step 4.   Add the partial products.

Multiply and check. Answers are on page 161.

1.
63	92	43	50	82	73	32
× 21	× 33	× 12	× 68	× 34	× 31	× 43

2.
421	723	504	612	334	522
× 34	× 13	× 22	× 41	× 21	× 14

3.
802	343	611	912	732	431
× 342	× 222	× 876	× 344	× 133	× 213

4.
5,203	9,332	4,112	6,330	8,231
× 123	× 331	× 144	× 221	× 132

Multiplication with Carrying

When you multiply two digits, the product is often a two-digit number. Inside a problem, you must **carry** the left digit to the next number you are multiplying. Add the digit you carry to the next product.

EXAMPLE: Multiply 76 × 9 =

$$\begin{array}{r} \overset{5}{7}6 \\ \times\ \ 9 \\ \hline 684 \end{array}$$

Step 1. 9 × 6 = 54. Write the 4 under the units column and carry the 5 to the tens column.

Step 2. 9 × 7 = 63. Add the 5 that you carried. 63 + 5 = 68.

Multiply and check. Answers are on page 161.

1.
$$\begin{array}{r} 42 \\ \times\ 8 \end{array} \quad \begin{array}{r} 65 \\ \times\ 7 \end{array} \quad \begin{array}{r} 58 \\ \times\ 9 \end{array} \quad \begin{array}{r} 78 \\ \times\ 4 \end{array} \quad \begin{array}{r} 97 \\ \times\ 3 \end{array} \quad \begin{array}{r} 66 \\ \times\ 2 \end{array} \quad \begin{array}{r} 19 \\ \times\ 6 \end{array}$$

2.
$$\begin{array}{r} 74 \\ \times\ 6 \end{array} \quad \begin{array}{r} 82 \\ \times\ 9 \end{array} \quad \begin{array}{r} 37 \\ \times\ 8 \end{array} \quad \begin{array}{r} 68 \\ \times\ 5 \end{array} \quad \begin{array}{r} 39 \\ \times\ 4 \end{array} \quad \begin{array}{r} 22 \\ \times\ 9 \end{array} \quad \begin{array}{r} 54 \\ \times\ 3 \end{array}$$

3.
$$\begin{array}{r} 45 \\ \times\ 9 \end{array} \quad \begin{array}{r} 38 \\ \times\ 3 \end{array} \quad \begin{array}{r} 84 \\ \times\ 8 \end{array} \quad \begin{array}{r} 43 \\ \times\ 7 \end{array} \quad \begin{array}{r} 92 \\ \times\ 6 \end{array} \quad \begin{array}{r} 76 \\ \times\ 5 \end{array} \quad \begin{array}{r} 55 \\ \times\ 7 \end{array}$$

4.
$$\begin{array}{r} 534 \\ \times\ 8 \end{array} \quad \begin{array}{r} 287 \\ \times\ 5 \end{array} \quad \begin{array}{r} 125 \\ \times\ 9 \end{array} \quad \begin{array}{r} 764 \\ \times\ 4 \end{array} \quad \begin{array}{r} 628 \\ \times\ 3 \end{array} \quad \begin{array}{r} 735 \\ \times\ 6 \end{array}$$

5.
$$\begin{array}{r} 772 \\ \times\ 6 \end{array} \quad \begin{array}{r} 679 \\ \times\ 3 \end{array} \quad \begin{array}{r} 374 \\ \times\ 5 \end{array} \quad \begin{array}{r} 726 \\ \times\ 4 \end{array} \quad \begin{array}{r} 663 \\ \times\ 8 \end{array} \quad \begin{array}{r} 929 \\ \times\ 7 \end{array}$$

6.
$$\begin{array}{r} 548 \\ \times\ 9 \end{array} \quad \begin{array}{r} 692 \\ \times\ 8 \end{array} \quad \begin{array}{r} 687 \\ \times\ 2 \end{array} \quad \begin{array}{r} 865 \\ \times\ 5 \end{array} \quad \begin{array}{r} 427 \\ \times\ 6 \end{array} \quad \begin{array}{r} 582 \\ \times\ 7 \end{array}$$

7.
36	42	25	84	77	64	33
× 38	× 65	× 43	× 36	× 46	× 55	× 78

8.
57	88	29	36	44	85	76
× 48	× 62	× 34	× 53	× 29	× 18	× 37

9.
54	85	99	28	67	59	87
× 63	× 47	× 38	× 19	× 32	× 43	× 26

Zero multiplied by any number is 0. When multiplying by zero, you can save time by putting a single 0 in the column where the multiplying begins. Compare these two examples.

ACCEPTABLE METHOD:
```
      42
  ×   30
      00
    1 26
   1,260
```

BETTER METHOD:
```
      42
  ×   30
   1,260
```

10.
48	75	92	34	91	83	54
× 30	× 50	× 60	× 40	× 80	× 20	× 70

11.
67	88	96	45	79	87	25
× 60	× 90	× 20	× 30	× 80	× 40	× 30

12.
86	63	73	68	82	76	53
× 70	× 90	× 40	× 50	× 80	× 20	× 60

13.

193	384	942	676	523	738
× 38	× 45	× 69	× 66	× 54	× 82

14.

866	679	492	536	458	966
× 24	× 34	× 73	× 57	× 65	× 83

15.

778	684	277	756	828	443
× 63	× 46	× 28	× 82	× 72	× 49

16.

872	183	384	342	648	375
× 267	× 492	× 288	× 595	× 876	× 416

17.

793	285	246	737	594	833
× 407	× 506	× 208	× 304	× 409	× 207

```
    793
×   407
  5 551
317 20
322,751
```

18.

4,623	9,127	5,194	8,386	4,388
× 47	× 36	× 52	× 85	× 63

19.

6,454	2,918	7,028	2,947	8,076
× 145	× 237	× 415	× 338	× 265

Horizontal Multiplication with Carrying

Sometimes a multiplication problem is written horizontally (with the numbers standing side by side). Rewrite the problem with units under units, tens under tens, and so on. To save time, put the smaller number below.

EXAMPLE: $386 \times 47 =$

ACCEPTABLE METHOD:

```
            47
    ×     386
          282
        3 76
      14 1
      18,142
```

BETTER METHOD:

```
        386
    ×    47
       2 702
      15 44
      18,142
```

Multiply and check. Answers are on page 162.

1. $526 \times 8 =$ $7 \times 473 =$ $609 \times 4 =$

2. $9 \times 733 =$ $487 \times 6 =$ $5 \times 389 =$

3. $40 \times 94 =$ $37 \times 70 =$ $90 \times 46 =$

4. $87 \times 25 =$ $36 \times 92 =$ $82 \times 47 =$

5. $49 \times 288 =$ $732 \times 86 =$ $75 \times 374 =$

6. $1,695 \times 46 =$ $38 \times 8,362 =$ $2,487 \times 56 =$

Multiplication by 10, 100, and 1000

You have already learned a shortcut for multiplying numbers by zero. Multiplying by 10, 100, and 1000 is even easier.

To multiply a number by 10, put one zero at the right of the number.

To multiply a number by 100, put two zeros at the right of the number.

To multiply a number by 1000, put three zeros at the right of the number.

EXAMPLE: $10 \times 82 = 820$
$100 \times 36 = 3,600$
$1000 \times 593 = 593,000$

Multiply and check. Answers are on page 162.

1. $10 \times 92 =$ $10 \times 136 =$ $10 \times 8 =$ $10 \times 204 =$

2. $87 \times 10 =$ $907 \times 10 =$ $460 \times 10 =$ $38 \times 10 =$

3. $100 \times 16 =$ $100 \times 288 =$ $100 \times 90 =$ $100 \times 43 =$

4. $7 \times 100 =$ $68 \times 100 =$ $493 \times 100 =$ $60 \times 100 =$

5. $1000 \times 17 =$ $1000 \times 9 =$ $1000 \times 208 =$ $1000 \times 46 =$

6. $123 \times 1000 =$ $14 \times 1000 =$ $785 \times 1000 =$ $3 \times 1000 =$

7. $10 \times 290 =$ $100 \times 400 =$ $20 \times 1000 =$ $360 \times 10 =$

Multiplication Applications

In these problems you'll have a chance to use your multiplication skills. Except for **product** and **times,** there are no key words that tell you to multiply. Instead of key words, watch for certain kinds of problems. For example, you may be told the price of one item and be asked to find the price of several items. Or you may be told the distance a car can travel in one hour and be asked to find how far it can travel in several hours. You may be told the weight of one thing and be asked to find the weight of several things. Each of these problems means to multiply.

Give every answer the correct label such as $ or feet. **Answers are on page 162.**

1. Jack makes $189.80 a week. There are 52 weeks in a year. How much does Jack make in a year?

2. One pound of ground beef costs $1.69. How much do four pounds of ground beef cost?

3. On the scale of a road map one inch equals 32 miles. How far apart are two cities which are 9 inches apart on the map?

4. Don drives an average of 17 miles on a gallon of gas when he drives on the highway. How many miles can he drive on the highway with 18 gallons of gas?

5. There are 12 inches in a foot. How many inches are there in 15 feet?

6. Jose earns $4.35 an hour. How much does he earn working 35 hours a week?

7. One case of canned beans weighs 8 pounds. How much do 36 cases weight?

8. Mark is paying back a loan. He pays $62.50 a month for 24 months. Find the amount he is paying back.

9. The electricity for running a T.V. costs 2¢ an hour. The Green family watches T.V. for an average of 43 hours each week. How much do they spend in a week to pay for the electricity to run their T.V.?

10. A dozen peaches cost $1.08. Find the cost of six dozen peaches.

11. Mr. Walek makes a profit of $2.30 for every shirt that he sells in his store. Last week he sold 156 shirts. What was his profit on the shirts?

12. Pat can type 89 words a minute. How many words can she type in 15 minutes?

13. Jeff drives at an average speed of 58 miles per hour on the highway. How far can he drive in four hours?

14. There are 16 ounces in a pound. How many ounces are there in 24 pounds?

Division Facts

The division facts are the opposite of the multiplication facts. If you know the multiplication facts, you should have no trouble with this exercise. Do this exercise without help. Check your answers on page 162. Then study the facts you missed until you can do this exercise with no errors.

1. $7\overline{)42}$ $3\overline{)3}$ $6\overline{)48}$ $5\overline{)30}$ $4\overline{)4}$ $5\overline{)0}$ $4\overline{)28}$

2. $8\overline{)24}$ $1\overline{)9}$ $3\overline{)12}$ $9\overline{)18}$ $9\overline{)54}$ $8\overline{)40}$ $5\overline{)35}$

3. $8\overline{)16}$ $6\overline{)6}$ $6\overline{)54}$ $2\overline{)8}$ $4\overline{)12}$ $4\overline{)24}$ $8\overline{)64}$

4. $3\overline{)0}$ $5\overline{)45}$ $7\overline{)21}$ $1\overline{)1}$ $2\overline{)18}$ $8\overline{)72}$ $2\overline{)12}$

5. $3\overline{)27}$ $1\overline{)4}$ $2\overline{)16}$ $6\overline{)42}$ $5\overline{)5}$ $2\overline{)14}$ $3\overline{)24}$

6. $5\overline{)15}$ $7\overline{)49}$ $1\overline{)7}$ $6\overline{)12}$ $8\overline{)32}$ $6\overline{)36}$ $7\overline{)56}$

7. $8\overline{)8}$ $5\overline{)40}$ $7\overline{)63}$ $6\overline{)0}$ $7\overline{)35}$ $1\overline{)5}$ $3\overline{)21}$

8. $1\overline{)6}$ $4\overline{)16}$ $5\overline{)20}$ $8\overline{)56}$ $9\overline{)72}$ $9\overline{)81}$ $8\overline{)48}$

9. $9\overline{)36}$ $3\overline{)9}$ $2\overline{)6}$ $1\overline{)8}$ $9\overline{)63}$ $6\overline{)30}$ $3\overline{)6}$

10. $2\overline{)2}$ $4\overline{)36}$ $3\overline{)18}$ $5\overline{)25}$ $7\overline{)14}$ $7\overline{)7}$ $6\overline{)24}$

11. $9\overline{)27}$ $8\overline{)0}$ $2\overline{)10}$ $7\overline{)28}$ $2\overline{)4}$ $5\overline{)10}$ $4\overline{)20}$

12. $4\overline{)8}$ $4\overline{)32}$ $3\overline{)15}$ $1\overline{)2}$ $9\overline{)0}$ $6\overline{)18}$ $9\overline{)45}$

Division by One Digit

The answer to a division problem is called the **quotient.** To find a quotient, repeat the four steps listed below until you complete the problem.

1. Divide.
2. Multiply.
3. Subtract and compare.
4. Bring down the next number.

EXAMPLE: Divide 6 into 504.

$$\frac{8}{6)\overline{504}}$$

Step 1. **Divide:** 6 goes into 50 8 times. Notice that 9 is too large and 7 is too small for the first step. $9 \times 6 = 54$, which is more than 50. $7 \times 6 = 42$, but $8 \times 6 = 48$, which is closer to 50. Write 8 above the tens place.

$$\begin{array}{r} 8 \\ 6)\overline{504} \\ 48 \end{array}$$

Step 2. **Multiply:** $8 \times 6 = 48$. Write 48 under 50.

$$\begin{array}{r} 8 \\ 6)\overline{504} \\ \underline{48} \\ 2 \end{array}$$

Step 3. **Subtract:** $50 - 48 = 2$. **Compare** to be sure that what you get by subtraction is less than what you divide by. 2 is less than 6.

$$\begin{array}{r} 8 \\ 6)\overline{504} \\ \underline{48} \\ 24 \end{array}$$

Step 4. **Bring down the next number:** 4.

$$\begin{array}{r} 84 \\ 6)\overline{504} \\ \underline{48} \\ 24 \end{array}$$

Step 5. **Divide:** 6 goes into 24 4 times. Write 4 above the units place.

$$\begin{array}{r} 84 \\ 6)\overline{504} \\ \underline{48} \\ 24 \\ 24 \end{array}$$

Step 6. **Multiply:** $4 \times 6 = 24$. Write 24 under 24.

$$\begin{array}{r} 84 \\ 6)\overline{504} \\ \underline{48} \\ 24 \\ \underline{24} \\ 0 \end{array}$$

Step 7. **Subtract:** $24 - 24 = 0$. **Compare:** 0 is less than 6.

Writing every step, as the example shows in step 7, is called **long division.** In **short division** you write only the answer and the number you get by subtracting.

EXAMPLE:

$$6\overline{)50^24}$$
with quotient 8 4

You can use either long division or short division when you divide by one digit.

To check a division problem, multiply your answer (the quotient) by the number you divided by. The product should equal the number you divided.

EXAMPLE:

$$6\overline{)504}$$ quotient 84

Check:

$$\begin{array}{r} 84 \\ \times\ \ 6 \\ \hline 504 \end{array}$$

Divide and check. Answers are on page 162.

1. $3\overline{)141}$ $9\overline{)207}$ $2\overline{)170}$ $5\overline{)280}$ $7\overline{)308}$

2. $8\overline{)616}$ $7\overline{)252}$ $6\overline{)270}$ $5\overline{)345}$ $4\overline{)348}$

3. $3\overline{)237}$ $6\overline{)348}$ $5\overline{)320}$ $8\overline{)704}$ $3\overline{)276}$

4. $9\overline{)2,484}$ $2\overline{)1,678}$ $7\overline{)2,562}$ $3\overline{)2,553}$ $6\overline{)5,568}$

5. $7\overline{)3,108}$ $8\overline{)1,864}$ $4\overline{)3,452}$ $6\overline{)5,628}$ $5\overline{)3,270}$

Division with Remainders

If you do not get zero in the last subtraction step of a division problem, you will have a **remainder.** To check a division problem with a remainder, multiply the answer of your division by the number you divided by and add the remainder.

```
EXAMPLE:      96 r 3     Check:      96
           4)387                  ×    4
              36                      384
              27                  +     3
              24                      387
               3
```

Divide and check. Answers are on page 162.

1. 7)292   2)79   9)204   6)316   3)236

2. 4)243   3)169   8)357   7)515   6)398

3. 5)467   9)698   8)300   2)199   7)519

4. 4)270   3)245   7)353   9)431   5)328

5. 3)274   8)566   6)257   5)338   4)307

6. 2)111   9)735   3)149   7)565   6)225

7. $8\overline{)1,542}$ $7\overline{)5,050}$ $9\overline{)2,743}$ $3\overline{)1,760}$

8. $6\overline{)3,845}$ $9\overline{)7,355}$ $8\overline{)7,366}$ $5\overline{)2,277}$

9. $8\overline{)2,183}$ $6\overline{)4,765}$ $2\overline{)1,937}$ $4\overline{)2,330}$

10. $9\overline{)2,810}$ $7\overline{)3,951}$ $5\overline{)2,709}$ $3\overline{)2,546}$

11. $5\overline{)4,561}$ $8\overline{)5,623}$ $6\overline{)2,317}$ $2\overline{)1,815}$

12. $3\overline{)2,567}$ $6\overline{)4,367}$ $7\overline{)3,085}$ $4\overline{)2,758}$

13. $2\overline{)1,231}$ $8\overline{)3,599}$ $9\overline{)6,357}$ $5\overline{)2,404}$

Division by Larger Numbers

To divide by two-digit and three-digit numbers, you must **estimate** how many times one number divides into another number. When you estimate, you make a guess.

EXAMPLE: Divide 62 into 2,976.

```
      48
62)2,976
   2 48
    496
    496
      0
```

Step 1. **Estimate** how many times 62 divides into 297. To do this, ask yourself how many times 6 divides into 29. 6 goes into 29 4 times. Write 4 above the 7.

Step 2. **Multiply:** $4 \times 62 = 248$. Write 248 under 297.

Step 3. **Subtract:** $297 - 248 = 49$. And **compare:** 49 is less than 62.

Step 4. **Bring down the next number:** 6.

Step 5. **Estimate** how many times 62 divides into 496. To do this, ask yourself how many times 6 goes into 49. 6 goes into 49 8 times. Write 8 above the 6.

Step 6. **Multiply:** $8 \times 62 = 496$. Write 496 under 496.

Step 7. **Subtract:** $496 - 496 = 0$.

Check:

```
       48
×      62
       96
    2 88
    2,976
```

Often your first estimate will be wrong. Work with a pencil and a good eraser. Decide if your estimate is too small or too large and try again.

Divide and check. Answers are on page 163.

1. 43)301   95)380   58)464   63)378

2. 24)192   89)623   38)228   62)310

3. 78)732   18)105   87)360   60)495

4. 41)1,968   67)3,752   52)3,380   49)2,548

5. 23)2,093   72)5,544   54)3,510   28)2,492

6. 37)1,406   79)3,634   41)1,517   66)3,894

7. 22)1,616   85)6,910   39)1,412   91)2,406

8. 86)2,554   44)2,216   76)6,237   29)2,132

9. 46)26,772   67)23,182   88)44,528   21)17,913

10. 72)49,782   25)17,613   96)36,210   77)32,822

11. 491)3,928   619)3,714   123)1,107   344)1,376

12. 264)1,848   527)3,162   605)5,445   268)2,144

13. 418)13,376   673)25,574   444)27,528   307)16,271

14. 641)30,327   787)43,435   283)10,308   512)31,145

15. 336)29,568   871)40,937   420)26,460   567)29,484

Horizontal Division

If a problem is written with the ÷ sign, rewrite the problem using the $\overline{)}$ sign. Notice that the numbers are turned around with these two signs.

EXAMPLE: 1,898 ÷ 73 =   *Change To:*

$$\begin{array}{r} 26 \\ 73\overline{)1,898} \\ 1\,46 \\ \hline 438 \\ 438 \\ \hline 0 \end{array}$$

Divide and check. Answers are on page 163.

1. 1,872 ÷ 8 =   2,508 ÷ 6 =   4,263 ÷ 7 =

2. 4,614 ÷ 9 =   1,554 ÷ 4 =   1,963 ÷ 4 =

3. 4,186 ÷ 46 =   4,650 ÷ 62 =   3,256 ÷ 88 =

4. 5,161 ÷ 97 =   2,259 ÷ 34 =   6,144 ÷ 72 =

5. 1,728 ÷ 18 =   4,368 ÷ 91 =   4,081 ÷ 53 =

Division Applications

With these problems you'll have a chance to use your division skills. Except for **quotient,** there are no key words that tell you to divide. Instead of key words, watch for certain kinds of problems. You may be told the price of several items and be asked to find the price of one item. Or you may be asked to find how many of a certain unit there are in a larger number of units. You may be asked to find an **average.** An average is a total divided by the number of items in the total.

Give every answer the correct label such as $ or ounces. **Answers are on page 163.**

1. How many two-pound boxes can be filled with 178 pounds of salt?

2. Last year Mr. and Mrs. Jones paid $2,760 in mortgage payments. There are 12 months in a year. How much did they pay each month on their mortgage?

3. Mark drives at an average speed of 52 miles per hour. How many hours will he need to drive 624 miles?

4. Nora and Doug bought a new T.V. and a stereo for $416.50. They agreed to pay in 17 equal monthly payments. How much will they pay each month?

5. There are 16 ounces in a pound. How many pounds are there in 560 ounces?

6. There are 7 people in the Howard family. Their total income last year was $12,950. What was the average yearly amount for each member of the family?

7. Silvia packs cans into boxes. Each box holds 16 cans. How many boxes can she fill with 3,776 cans?

8. Carl is a truck driver. In 15 work days he drove a total of 7,305 miles. What was the average distance he drove each day?

9. There are 1,760 yards in a mile. How many miles are there in 40,480 yards?

10. On the scale of a road map one inch is equal to 45 miles. How far apart on the map are two cities that are really 495 miles apart?

11. One dozen apples cost 84¢. How much does one apple cost? (one dozen = 12)

12. A train travels at an average speed of 42 miles per hour. How many hours will it take the train to travel 1,134 miles?

13. Celeste sells shoes on commission. In the last three weeks she made $370.80 in commission. What was the average amount she made each week?

Whole Number Review

These problems will help you find out if you need to review the whole number section of this book. When you finish, look at the chart to see which pages you should review.

In problems 1 to 4 read each number. Then write out the missing word in the name of the number.

1. 18,200   eighteen _____, two hundred.

2. 60,009,040   sixty _____, nine_____, forty.

3. 5,300,800   five_____, three hundred_____,

   eight_____.

4. 72,090,030   seventy-two_____, ninety_____, thirty.

In problems 5 to 8, read the numbers. Then write the numbers in figures.

5. five hundred forty _____

6. fifteen thousand, two hundred six _____

7. four million, one hundred twenty thousand, eight_____

8. ninety million, seventy-six thousand, eight hundred_____

9.   478
      + 321

10.   19,452
       + 30,136

11. 425 + 34 =

12. 732 + 9,255 =

13.   86
       + 75

14.   43
       96
       + 77

15.   3,487
       296
       6,755
       + 863

16.   10,926
       4,387
       32,664
       + 1,955

17. 6,927 + 434 + 56 =

18. 34,209 + 876 + 1,653 =

19. Pete bought a shirt for $12.69, a pair of jeans for $16.88, and a pair of socks for $1.75. The sales tax was $2.51. What was his total bill including tax?

20. At the Municipal Power Plant 63 employees work from 7 A.M. to 3 P.M. 46 employees work from 3 P.M. to 11 P.M. 39 employees work from 11 P.M. to 7 A.M. Find the total number of employees at the plant.

21.
```
   87
 − 34
```

22.
```
  573
 − 251
```

23.
```
  87,274
 − 25,142
```

24.
```
   83
 −  9
```

25.
```
  653
 − 287
```

26.
```
  41,382
 − 19,187
```

27.
```
  503
 − 274
```

28.
```
  60,300
 − 28,177
```

29. 800 − 73 =

30. 12,603 − 9,258 =

31. 90,000 − 4,782 =

32. 50,030 − 8,916 =

33. John bought a camera on sale for $68.50. Before the sale, the camera cost $85. How much did John save by buying the camera on sale?

34. 403 people belong to the Midvale City Employees' Union. 287 of the members voted to strike. How many members did not vote to strike?

35.   84
    × 2

36.   512
    × 4

37.   62
    × 34

38.   912
    × 231

39.   48
    × 9

40.   73
    × 64

41.   83
    × 70

42.   536
    × 273

43. 94 × 60 =

44. 26 × 785 =

45. 2,706 × 75 =

46. 39 × 4,086 =

47. 10 × 123 =

48. 16 × 100 =

49. 1000 × 9 =

50. One pound of roast beef costs $1.89. How much do five pounds of roast beef cost?

51. There are 5,280 feet in a mile. How many feet are there in 17 miles?

52. $8\overline{)584}$ **53.** $4\overline{)196}$ **54.** $9\overline{)779}$ **55.** $6\overline{)298}$

56. $7\overline{)3,366}$ **57.** $52\overline{)4,836}$ **58.** $39\overline{)24,463}$ **59.** $347\overline{)18,738}$

60. $3,960 \div 8 =$ **61.** $7,224 \div 9 =$

62. $4,662 \div 63 =$ **63.** $3,627 \div 42 =$

64. Colin drives at an average speed of 48 miles per hour. How many hours does he need to drive 768 miles?

65. Nick makes a $45 payment every month for his used car. How many months will it take Nick to pay the $1,620 he owes?

Check your answers on page 163. Then turn to the review pages for the problems you missed. Correct your answers before going on to the next page.

If you missed problems	review pages
1 to 8	7 to 8
9 to 12	9 to 11
13 to 18	12 to 15
19 to 20	16 to 17
21 to 23	18 to 19
24 to 32	20 to 24
33 to 34	25 to 26
35 to 38	27 to 31
39 to 46	32 to 35
47 to 49	36
50 to 51	37 to 38
52 to 57	39 to 43
58 to 63	44 to 47
64 to 65	48 to 49

Step One to Fraction Skill

These problems will help you find out if you need work in the fraction section of this book. Do all the problems you can. When you are finished, look at the chart to see which page you should go to next.

1. For each picture write a fraction that shows what part of the picture is shaded.

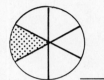 _____ _____ _____ 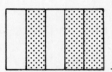 _____

2. There are 12 months in a year.
7 months are what fraction of a year?

3. There are 20 nickels in a dollar.
3 nickels are what fraction of a dollar?

4. Circle the proper fractions in this list:

$$\frac{9}{3} \qquad \frac{3}{4} \qquad \frac{8}{2} \qquad \frac{5}{5} \qquad 2\frac{3}{5} \qquad \frac{5}{12}$$

5. Circle the improper fractions in this list:

$$9\frac{4}{7} \qquad \frac{3}{8} \qquad \frac{8}{3} \qquad \frac{2}{2} \qquad \frac{7}{3} \qquad \frac{6}{9}$$

6. Circle the mixed numbers in this list:

$$\frac{3}{10} \qquad \frac{1}{2} \qquad 6\frac{2}{5} \qquad \frac{4}{4} \qquad \frac{8}{9} \qquad \frac{10}{3}$$

7. Reduce each fraction to lowest terms.

$$\frac{6}{54} = \qquad \frac{36}{48} = \qquad \frac{15}{65} = \qquad \frac{60}{900} =$$

8. Raise each fraction to higher terms by finding the missing number.

$$\frac{3}{8} = \frac{}{56} \qquad \frac{6}{25} = \frac{}{75} \qquad \frac{5}{12} = \frac{}{48} \qquad \frac{7}{10} = \frac{}{100}$$

9. Change each improper fraction to a whole or mixed number. Reduce each fraction that is left.

$$\frac{24}{10} = \qquad \frac{39}{9} = \qquad \frac{15}{15} = \qquad \frac{20}{3} =$$

10. Change each mixed number to an improper fraction.

$5\frac{2}{3} =$ $3\frac{3}{7} =$ $4\frac{2}{5} =$ $8\frac{1}{5} =$

11. Circle the bigger fraction in each pair.

$\frac{7}{12}$ or $\frac{5}{8}$ $\frac{1}{4}$ or $\frac{3}{10}$ $\frac{5}{6}$ or $\frac{3}{4}$ $\frac{8}{15}$ or $\frac{2}{3}$

12. Mr. and Mrs. Green bought a $35,000 house. They made a down payment of $7,000. The down payment was what fraction of the price of the house?

13. There are 45 members in the Oakdale Tenants' Group. 25 members went to the last meeting. What fraction of the members went to the meeting?

14.
$$\begin{array}{r} \frac{2}{5} \\ + \frac{1}{5} \\ \hline \end{array}$$

15.
$$\begin{array}{r} \frac{3}{8} \\ + \frac{3}{8} \\ \hline \end{array}$$

16.
$$\begin{array}{r} 3\frac{9}{16} \\ + 2\frac{11}{16} \\ \hline \end{array}$$

17.
$$\begin{array}{r} \frac{5}{9} \\ + \frac{2}{3} \\ \hline \end{array}$$

18.
$$\begin{array}{r} \frac{3}{7} \\ + \frac{1}{2} \\ \hline \end{array}$$

19.
$$\begin{array}{r} 3\frac{3}{8} \\ 5\frac{1}{4} \\ + 2\frac{5}{12} \\ \hline \end{array}$$

20. $3\frac{1}{2} + 1\frac{3}{4} + 2\frac{7}{10} =$

21. $4\frac{1}{3} + 3\frac{3}{20} + 2\frac{5}{6} =$

22. Jane is $64\frac{1}{2}$ inches tall. Her husband Ron is $2\frac{7}{8}$ inches taller. How tall is Ron?

23. Joe bought $3\frac{5}{16}$ pounds of nails, $1\frac{1}{2}$ pounds of wood screws, and $\frac{7}{8}$ pound of staples. What was the weight of the things Joe bought?

24.
$$\begin{array}{r} \frac{9}{10} \\ - \frac{7}{10} \\ \hline \end{array}$$

25.
$$\begin{array}{r} \frac{15}{16} \\ - \frac{11}{16} \\ \hline \end{array}$$

26.
$$\begin{array}{r} \frac{5}{8} \\ - \frac{1}{3} \\ \hline \end{array}$$

27.
$$\begin{array}{r} 6 \\ - 2\frac{5}{9} \\ \hline \end{array}$$

28. $8\dfrac{1}{5}$
           $-2\dfrac{4}{5}$

29. $9\dfrac{1}{4}$
           $-6\dfrac{7}{8}$

30. $7\dfrac{4}{9} - 4\dfrac{5}{6} =$

31. $8\dfrac{1}{3} - 4\dfrac{3}{5} =$

32. Carlos usually works 40 hours a week. Last week he was sick and missed $15\dfrac{3}{4}$ hours of work. How many hours did Carlos work last week?

33. In January John weighed $180\dfrac{1}{2}$ pounds. In June he weighed $158\dfrac{7}{8}$ pounds. How much weight did he lose?

34. $\dfrac{5}{9} \times \dfrac{2}{3} =$

35. $\dfrac{4}{15} \times \dfrac{25}{28} =$

36. $\dfrac{9}{10} \times \dfrac{8}{9} =$

37. $\dfrac{5}{9} \times 12 =$

38. $15 \times 4\dfrac{2}{3} =$

39. $\dfrac{3}{5} \times 1\dfrac{1}{9} =$

40. $2\dfrac{1}{4} \times 4\dfrac{2}{3} =$

41. $1\dfrac{1}{8} \times 5\dfrac{1}{3} =$

42. Lois bought $3\dfrac{1}{4}$ pounds of chicken for $1.20 a pound. How much did she pay for the chicken?

43. Jack and Karen want to buy a $36,000 house. They have to make a down payment of $\dfrac{1}{5}$ of the price of the house. How much is the down payment?

44. $\dfrac{4}{5} \div \dfrac{8}{15} =$

45. $8 \div \dfrac{6}{7} =$

46. $4\dfrac{1}{6} \div \dfrac{5}{9} =$

47. $\dfrac{5}{12} \div 10 =$

48. $\dfrac{5}{8} \div 2\dfrac{3}{4} =$ **49.** $6 \div 4\dfrac{1}{2} =$ **50.** $3\dfrac{1}{2} \div 2\dfrac{4}{5} =$ **51.** $2\dfrac{2}{5} \div 2\dfrac{1}{10} =$

52. Janet paid $4.55 for $3\dfrac{1}{2}$ pounds of fish. How much did one pound of fish cost?

53. Richard cut a board $60\dfrac{3}{4}$ inches long into three equal pieces. How long was each piece?

Check your answers on page 164. Then complete the chart below.

Problem numbers	Number of problems in this section	Number of problems you got right in this section	
1 to 13	13	_____	If you had fewer than 10 problems right, go to page 59.
14 to 23	10	_____	If you had fewer than 8 problems right, go to page 70.
24 to 33	10	_____	If you had fewer than 8 problems right, go to page 78.
34 to 43	10	_____	If you had fewer than 8 problems right, go to page 85.
44 to 53	10	_____	If you had fewer than 8 problems right, go to page 93.

If you missed no more than 12 problems, correct them and go to Step One to Decimal Skill on page 105.

Writing Fractions

A fraction is two numbers that show a part of some whole. Three dimes are three of the ten equal parts of one dollar. Three dimes are $\frac{3}{10}$ or three tenths of a dollar. Five ounces are five of the sixteen equal parts of one pound. Five ounces are $\frac{5}{16}$ or five sixteenths of a pound.

The top number of a fraction is called the **numerator.** The numerator tells you how many parts you have. The bottom number is called the **denominator.** The denominator tells you how many parts are in the whole.

In the fraction $\frac{3}{10}$, 3 is the numerator and 10 is the denominator. You have 3 parts. The whole has 10 parts. In the fraction $\frac{5}{16}$, 5 is the numerator and 16 is the denominator. You have 5 parts. The whole has 16 parts.

The pictures below are partly shaded. The fraction beside each picture tells what part of the picture is shaded. The numerator (top number) tells how many parts are shaded. The denominator (bottom number) tells how many parts are in the whole picture.

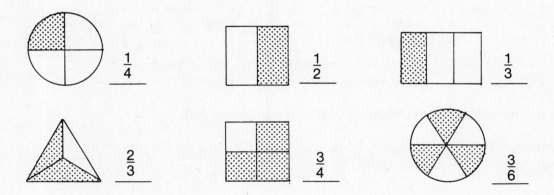

For each picture write a fraction that tells what part of the picture is shaded. Answers are on page 165.

1.

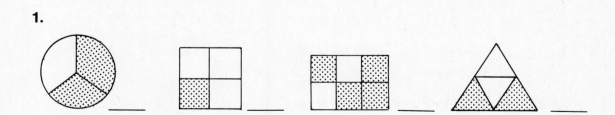

2.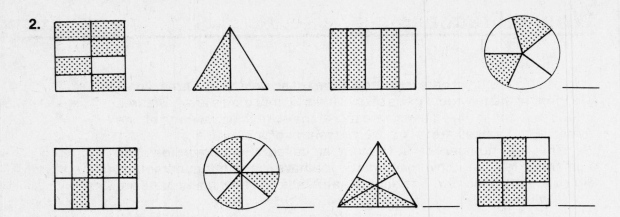

It is easy to write a fraction from words. First choose a denominator that tells the number of parts in the whole. Then choose a numerator that tells how many parts you have.

EXAMPLE: There are 12 inches in a foot. 11 inches are what fraction of a foot?

$\frac{11}{12}$ *Step 1.*   Choose the denominator. 12 tells how many parts are in one whole foot. 12 is the denominator.

         Step 2.   Choose the numerator. 11 tells how many parts you have. 11 is the numerator.

Write a fraction for each problem. Answers are on page 165.

4. There are 4 quarts in a gallon. Three quarts are what fraction of a gallon?

5. There are 10 dimes in a dollar. 7 dimes are what fraction of a dollar?

6. There are 60 minutes in an hour. 21 minutes are what fraction of an hour?

7. There are 1000 grams in a kilogram. 127 grams are what fraction of a kilogram?

8. There are 36 inches in a yard. 19 inches are what fraction of a yard?

9. There are 7 days in a week. Two days are what fraction of a week?

Identifying Types of Fractions

There are three types of fractions.

1. **Proper fraction:** The numerator (top number) is always less than the denominator. Examples: $\frac{5}{6}$, $\frac{2}{4}$, $\frac{19}{20}$. A proper fraction does not have all the parts of the whole. The value of a proper fraction is less than one whole.

2. **Improper fraction:** The numerator is as big or bigger than the denominator. Examples: $\frac{3}{2}$, $\frac{50}{40}$, $\frac{8}{8}$. When the numerator is bigger than the denominator, the improper fraction has a value of more than one whole. $\frac{3}{2}$ and $\frac{50}{40}$ are each more than one whole. When the numerator is equal to the denominator, the improper fraction is equal to one whole. $\frac{8}{8}$ is equal to one whole.

3. **Mixed number:** A whole number and a proper fraction are written next to each other. Examples: $9\frac{2}{3}$, $1\frac{1}{2}$, $10\frac{4}{15}$. A mixed number always has a value of more than one whole.

Answer the following questions. Answers are on page 165.

1. Circle the proper fractions in this list:

   $3\frac{1}{2}$ $\frac{6}{7}$ $\frac{8}{3}$ $\frac{4}{5}$ $\frac{5}{4}$ $\frac{2}{2}$ $9\frac{5}{7}$ $\frac{10}{10}$ $\frac{8}{200}$ $\frac{17}{4}$

2. Circle the improper fractions in this list:

   $\frac{19}{5}$ $\frac{5}{12}$ $8\frac{5}{6}$ $\frac{9}{10}$ $\frac{12}{9}$ $4\frac{2}{5}$ $\frac{15}{15}$ $\frac{8}{30}$ $\frac{24}{4}$ $\frac{2}{6}$

3. Circle the mixed numbers in this list:

   $\frac{9}{15}$ $8\frac{4}{7}$ $\frac{16}{15}$ $2\frac{3}{20}$ $\frac{19}{24}$ $\frac{24}{19}$ $3\frac{8}{9}$ $\frac{1}{500}$ $\frac{27}{3}$ $10\frac{1}{2}$

4. Circle the numbers with a value of more than one in this list:

   $\frac{4}{9}$ $\frac{12}{2}$ $6\frac{3}{5}$ $\frac{8}{8}$ $\frac{20}{7}$ $2\frac{9}{16}$ $3\frac{5}{8}$ $\frac{41}{4}$ $\frac{5}{24}$ $\frac{7}{7}$

Reducing

Reducing a fraction means dividing both the numerator and the denominator (the top and the bottom) by a number that goes into them evenly. Reducing changes the numbers in a fraction, but reducing does not change the value of a fraction. Remember that $\frac{1}{2}$ dollar or a 50¢ piece is the same amount of money as $\frac{2}{4}$ dollar or two 25¢ pieces.

EXAMPLE: Reduce $\frac{18}{20}$

$\frac{18 \div 2}{20 \div 2} = \frac{9}{10}$ *Step 1.* Divide both 18 and 20 by a number that goes evenly into both of them. 2 divides evenly into both 18 and 20.

$\frac{18}{20} = \frac{9}{10}$ *Step 2.* Check to see if another number, besides 1, goes evenly into 9 and 10. No other number divides evenly into both. The fraction $\frac{9}{10}$ is reduced as far as it will go.

The equal sign (=) tells you that $\frac{9}{10}$ has exactly the same value as $\frac{18}{20}$.

Sometimes a fraction can be reduced more than once.

EXAMPLE: Reduce $\frac{32}{48}$

$\frac{32 \div 8}{48 \div 8} = \frac{4}{6}$ *Step 1.* Divide both 32 and 48 by a number that goes evenly into both of them. 8 divides evenly into both 32 and 48.

$\frac{4 \div 2}{6 \div 2} = \frac{2}{3}$ *Step 2.* Check to see if another number goes evenly into 4 and 6. 2 divides evenly into both numbers. Divide 4 and 6 by 2.

$\frac{32}{48} = \frac{4}{6} = \frac{2}{3}$ *Step 3.* Check to see if another number goes evenly into 2 and 3. No other number divides evenly into both. $\frac{2}{3}$ is reduced as far as it will go.

A fraction that is reduced as far as it will go is in **lowest terms**.

When you reduce a fraction, be sure to:
1. Divide both the top and the bottom by the same number.
2. Check to see if the fraction can be reduced more.

Reduce each fraction to lowest terms. Answers are on page 165.

1. $\dfrac{9}{45} =$ $\dfrac{8}{72} =$ $\dfrac{6}{42} =$ $\dfrac{7}{56} =$ $\dfrac{5}{30} =$

2. $\dfrac{30}{65} =$ $\dfrac{25}{80} =$ $\dfrac{15}{45} =$ $\dfrac{35}{50} =$ $\dfrac{40}{55} =$

3. $\dfrac{80}{100} =$ $\dfrac{30}{60} =$ $\dfrac{40}{200} =$ $\dfrac{60}{600} =$ $\dfrac{20}{500} =$

4. $\dfrac{35}{42} =$ $\dfrac{32}{40} =$ $\dfrac{20}{32} =$ $\dfrac{18}{45} =$ $\dfrac{12}{30} =$

5. $\dfrac{9}{33} =$ $\dfrac{22}{24} =$ $\dfrac{36}{48} =$ $\dfrac{25}{45} =$ $\dfrac{56}{84} =$

6. $\dfrac{36}{63} =$ $\dfrac{24}{40} =$ $\dfrac{15}{27} =$ $\dfrac{13}{26} =$ $\dfrac{8}{96} =$

7. $\dfrac{16}{64} =$ $\dfrac{5}{105} =$ $\dfrac{70}{110} =$ $\dfrac{28}{32} =$ $\dfrac{49}{63} =$

8. $\dfrac{45}{60} =$ $\dfrac{36}{72} =$ $\dfrac{24}{30} =$ $\dfrac{44}{55} =$ $\dfrac{60}{140} =$

9. $\dfrac{48}{56} =$ $\dfrac{48}{60} =$ $\dfrac{12}{28} =$ $\dfrac{16}{32} =$ $\dfrac{90}{130} =$

Raising Fractions to Higher Terms

Later, when you add and subtract fractions, you will often need to raise fractions to higher terms. Raising to higher terms is the opposite of reducing. To raise a fraction to higher terms, multiply both the numerator and the denominator by the same number.

In the practice problems on this page, a new denominator is given for each fraction.

EXAMPLE: Raise the fraction to higher terms by finding the missing numerator. $\frac{3}{8} = \frac{}{40}$

$8\overline{)40}$ with quotient 5

Step 1. Divide the new denominator by the old denominator.

$\frac{3 \times 5}{8 \times 5} = \frac{15}{40}$

Step 2. Multiply both the old numerator and the old denominator by 5. $\frac{15}{40}$ is $\frac{3}{8}$ raised to higher terms.

To check a fraction you have raised to higher terms, reduce it. Reduce $\frac{15}{40}$ by 5. $\frac{15 \div 5}{40 \div 5} = \frac{3}{8}$. $\frac{3}{8}$ was the old fraction.

Raise each fraction to higher terms by finding the missing numerator. Answers are on page 165.

1. $\frac{3}{4} = \frac{}{24}$ $\frac{2}{9} = \frac{}{36}$ $\frac{7}{10} = \frac{}{80}$ $\frac{2}{5} = \frac{}{35}$ $\frac{5}{8} = \frac{}{40}$

2. $\frac{6}{7} = \frac{}{63}$ $\frac{1}{12} = \frac{}{36}$ $\frac{3}{5} = \frac{}{50}$ $\frac{4}{9} = \frac{}{54}$ $\frac{8}{11} = \frac{}{22}$

3. $\frac{2}{3} = \frac{}{36}$ $\frac{2}{9} = \frac{}{72}$ $\frac{7}{10} = \frac{}{60}$ $\frac{1}{8} = \frac{}{56}$ $\frac{4}{5} = \frac{}{45}$

4. $\frac{3}{4} = \frac{}{32}$ $\frac{1}{15} = \frac{}{45}$ $\frac{6}{11} = \frac{}{55}$ $\frac{2}{3} = \frac{}{60}$ $\frac{9}{10} = \frac{}{40}$

Changing Improper Fractions to Whole or Mixed Numbers

The answers to many fraction problems are improper fractions. These answers are easier to read if you change them to whole numbers or mixed numbers. In an improper fraction the numerator (top number) is as big or bigger than the denominator (bottom number). To change an improper fraction, divide the denominator into the numerator.

EXAMPLE: Change $\frac{30}{4}$ to a mixed number.

$$4\overline{)30} \quad \begin{array}{c}7\\28\\\hline 2\end{array}$$

Step 1. Divide the denominator into the numerator.

$\frac{30}{4} = 7\frac{2}{4}$ *Step 2.* Write the remainder over the old denominator. $\frac{2}{4}$

$7\frac{2}{4} = 7\frac{1}{2}$ *Step 3.* Reduce the fraction by 2. $7 \quad \frac{2 \div 2}{4 \div 2} = 7\frac{1}{2}$

EXAMPLE: Change $\frac{36}{9}$ to a whole number or a mixed number.

$$9\overline{)36} \quad 4$$

Step 1. Divide the denominator into the numerator.

Step 2. There is no remainder. The answer is 4.

Change each improper fraction to a whole number or a mixed number. Reduce each fraction that is left. Answers are on page 165.

1. $\frac{5}{2} =$ $\frac{13}{4} =$ $\frac{17}{3} =$ $\frac{43}{5} =$ $\frac{40}{9} =$

2. $\frac{37}{10} =$ $\frac{24}{4} =$ $\frac{18}{7} =$ $\frac{53}{9} =$ $\frac{56}{7} =$

3. $\frac{22}{12} =$ $\frac{41}{12} =$ $\frac{30}{6} =$ $\frac{22}{8} =$ $\frac{23}{7} =$

4. $\frac{48}{9} =$ $\frac{52}{8} =$ $\frac{77}{8} =$ $\frac{40}{16} =$ $\frac{29}{6} =$

Changing Mixed Numbers to Improper Fractions

In many multiplication and division of fractions problems, you will need to change mixed numbers into improper fractions. To change a mixed number to an improper fraction, follow these steps:

1. Multiply the denominator (bottom number) by the whole number.
2. Add the numerator.
3. Write the total over the denominator.

EXAMPLE: Change $7\frac{5}{8}$ to an improper fraction.

$7\frac{5}{8} = \frac{61}{8}$

Step 1. Multiply the denominator, 8, by the whole number, 7. $8 \times 7 = 56$

Step 2. Add the numerator. $56 + 5 = 61$

Step 3. Write the total, 61, over the denominator, 8. $\frac{61}{8}$

To check an improper fraction, change it back to a mixed number.

$$\begin{array}{r} 7\frac{5}{8} \\ 8\overline{)61} \\ \underline{56} \\ 5 \end{array}$$

Change each mixed number to an improper fraction. Answers are on page 165.

1. $6\frac{2}{3} =$ $3\frac{1}{2} =$ $5\frac{3}{4} =$ $2\frac{7}{10} =$ $1\frac{1}{6} =$

2. $3\frac{5}{8} =$ $9\frac{2}{3} =$ $4\frac{2}{7} =$ $6\frac{5}{6} =$ $1\frac{9}{10} =$

3. $8\frac{1}{3} =$ $2\frac{5}{12} =$ $9\frac{1}{4} =$ $4\frac{7}{8} =$ $3\frac{6}{7} =$

4. $7\frac{1}{2} =$ $1\frac{3}{16} =$ $4\frac{2}{9} =$ $3\frac{7}{12} =$ $4\frac{4}{5} =$

Comparing Fractions

To compare the size of two proper fractions, decide which fraction is closer to one whole in value. It is sometimes hard to compare the size of two fractions when both the numerators and the denominators are different. When the numerators and denominators are different, find a **common denominator** for both fractions. A common denominator is a number that both denominators divide into evenly.

EXAMPLE: Which fraction is bigger, $\frac{3}{5}$ or $\frac{5}{7}$?

Step 1. Find a common denominator for 5 and 7. The new denominator should be the lowest number that both 5 and 7 divide into evenly. To find the common denominator, keep multiplying the largest denominator, 7, $(7 \times 1, 7 \times 2,$ etc.) until you find the number that both denominators will go into evenly. $7 \times 5 = 35$. Both 5 and 7 divide into 35 evenly.

Step 2. Raise each fraction to a new fraction with a denominator of 35.

$$\frac{3 \times 7}{5 \times 7} = \frac{21}{35}, \quad \frac{5 \times 5}{7 \times 5} = \frac{25}{35}$$

Step 3. Decide which fraction is bigger. $\frac{25}{35}$ has 4 more 35ths than $\frac{21}{35}$. $\frac{25}{35}$ is bigger. $\frac{25}{35}$ is equal to $\frac{5}{7}$. $\frac{5}{7}$ is the bigger fraction.

EXAMPLE: Which fraction is bigger, $\frac{2}{5}$ or $\frac{7}{20}$?

Step 1. Find a common denominator for 5 and 20. The lowest number that both 5 and 20 divide into evenly is 20.

Step 2. Raise $\frac{2}{5}$ to a new fraction with a denominator of 20.

$$\frac{2 \times 4}{5 \times 4} = \frac{8}{20}$$

Step 3. Decide which fraction is bigger. $\frac{8}{20}$ has 1 more 20th than $\frac{7}{20}$. $\frac{8}{20}$ is bigger. $\frac{8}{20}$ is equal to $\frac{2}{5}$. $\frac{2}{5}$ is the bigger fraction.

Circle the bigger fraction in each pair. Answers are on page 166.

1. $\frac{1}{2}$ or $\frac{3}{5}$ $\frac{5}{12}$ or $\frac{1}{3}$ $\frac{1}{2}$ or $\frac{8}{15}$ $\frac{7}{9}$ or $\frac{3}{4}$

2. $\frac{4}{5}$ or $\frac{21}{25}$ $\frac{4}{7}$ or $\frac{1}{2}$ $\frac{1}{4}$ or $\frac{1}{6}$ $\frac{8}{15}$ or $\frac{2}{3}$

3. $\frac{2}{5}$ or $\frac{3}{8}$ $\frac{1}{3}$ or $\frac{5}{9}$ $\frac{7}{10}$ or $\frac{3}{4}$ $\frac{5}{16}$ or $\frac{3}{8}$

4. $\frac{7}{9}$ or $\frac{5}{6}$ $\frac{11}{17}$ or $\frac{2}{3}$ $\frac{5}{8}$ or $\frac{7}{12}$ $\frac{5}{6}$ or $\frac{7}{8}$

To compare three fractions, find a common denominator for all three.

EXAMPLE: Which fraction is biggest, $\frac{7}{12}$, $\frac{2}{3}$, or $\frac{5}{8}$?

Step 1.   Find a common denominator for 12, 3, and 8. The lowest number that 12, 3, and 8 divide into evenly is 24.

Step 2.   Raise each fraction to a new fraction with 24 as the denominator.

$$\frac{7 \times 2}{12 \times 2} = \frac{14}{24} \quad \frac{2 \times 8}{3 \times 8} = \frac{16}{24} \quad \frac{5 \times 3}{8 \times 3} = \frac{15}{24}$$

Step 3.   Decide which fraction is biggest. $\frac{16}{24}$ has 1 more 24th than $\frac{15}{24}$ and it has 2 more 24ths than $\frac{14}{24}$. $\frac{16}{24}$ is biggest. $\frac{16}{24}$ is equal to $\frac{2}{3}$. $\frac{2}{3}$ is the biggest fraction.

Circle the biggest fraction in each group. Answers are on page 166.

5. $\frac{1}{4}$, $\frac{3}{8}$, or $\frac{5}{16}$ $\frac{5}{12}$, $\frac{1}{4}$, or $\frac{5}{24}$ $\frac{3}{5}$, $\frac{1}{2}$, or $\frac{3}{4}$

6. $\frac{3}{4}$, $\frac{7}{9}$, or $\frac{29}{36}$ $\frac{9}{20}$, $\frac{3}{10}$, or $\frac{2}{5}$ $\frac{5}{9}$, $\frac{2}{3}$, or $\frac{11}{18}$

Finding What Part One Number Is of Another

To find what part one number is of another, make a fraction with the **part** as the numerator and the **whole** as the denominator. Then reduce the fraction to lowest terms.

EXAMPLE: George wants to save $1,600 to buy a motorcycle. So far he has saved $600. What fraction of the amount has he saved?

$$\frac{600 \div 200}{1,600 \div 200} = \frac{3}{8}$$

Step 1. Find the whole. The $1,600 price of the motorcycle is the whole. 1,600 is the denominator.

Step 2. Find the part. The $600 George has saved is the part. 600 is the numerator.

Step 3. Reduce the fraction by 200.

Write a fraction for each problem. Reduce each fraction to lowest terms. Answers are on page 166.

1. Mr. and Mrs. Smith make $840 a month. They spend $210 a month for rent. What fraction of their income goes for rent?

2. Jason took a test with 60 problems. He got 48 problems right. What fraction of the problems did he get right?

3. Shirley weighed 140 pounds. She went on a diet and lost 20 pounds. What fraction of her weight did she lose?

4. Jeff and Debbie bought new furniture for $450. They made a down payment of $100. The down payment was what fraction of the price of the furniture?

5. The Rialto Theater has 320 seats. On Saturday night 40 seats were empty. What fraction of the seats were empty?

6. There are 250 employees at the Midvale Hospital. 75 of the employees are men. What fraction of the employees are men?

Addition of Fractions with the Same Denominators

It is easy to add fractions with the same denominators (bottom numbers). Add the numerators (top numbers). Then write the total over the denominator. With mixed numbers, add the fractions and the whole numbers separately.

EXAMPLE: Add $3\frac{2}{5} + 4\frac{1}{5} =$

$$3\frac{2}{5}$$
$$+4\frac{1}{5}$$
$$\overline{7\frac{3}{5}}$$

Step 1. Add the numerators. $2 + 1 = 3$

Step 2. Write the total, 3, over the denominator.

Step 3. Add the whole numbers: $3 + 4 = 7$. The answer is $7\frac{3}{5}$.

Add each problem. Answers are on page 166.

1. $\frac{3}{7}$ $\frac{4}{6}$ $2\frac{4}{9}$ $6\frac{3}{8}$ $3\frac{7}{10}$
    $+\frac{2}{7}$ $+\frac{1}{6}$ $+8\frac{3}{9}$ $+4\frac{2}{8}$ $+5\frac{2}{10}$

2. $\frac{9}{20}$ $\frac{1}{5}$ $7\frac{1}{4}$ $3\frac{3}{16}$ $9\frac{5}{24}$
    $+\frac{4}{20}$ $+\frac{2}{5}$ $+4\frac{2}{4}$ $+9\frac{6}{16}$ $+5\frac{8}{24}$

3. $\frac{6}{18}$ $\frac{2}{11}$ $6\frac{4}{15}$ $4\frac{11}{32}$ $7\frac{17}{30}$
    $+\frac{11}{18}$ $+\frac{4}{11}$ $+6\frac{3}{15}$ $+8\frac{14}{32}$ $+9\frac{6}{30}$

Often the total in an addition of fractions problem can be reduced.

EXAMPLE: Add $\frac{5}{9} + \frac{1}{9} =$

$$\frac{5}{9}$$
$$+\frac{1}{9}$$
$$\overline{\frac{6}{9}} = \frac{2}{3}$$

Step 1. Add the numerators. $5 + 1 = 6$

Step 2. Write the total, 6, over the denominator.

Step 3. Reduce the answer, $\frac{6}{9}$, by 3. $\frac{6 \div 3}{9 \div 3} = \frac{2}{3}$

Always check an answer to find out if it can be reduced. A problem is not complete until the answer is reduced.

Add and reduce. Answers are on page 166.

4. $\frac{1}{6}$ $\frac{5}{12}$ $5\frac{1}{4}$ $8\frac{2}{10}$ $6\frac{9}{20}$
      $+\frac{1}{6}$ $+\frac{1}{12}$ $+2\frac{1}{4}$ $+3\frac{3}{10}$ $+4\frac{3}{20}$

5. $\frac{1}{10}$ $\frac{3}{8}$ $3\frac{13}{24}$ $8\frac{5}{16}$ $2\frac{17}{50}$
      $+\frac{7}{10}$ $+\frac{3}{8}$ $+9\frac{7}{24}$ $+5\frac{7}{16}$ $+9\frac{18}{50}$

6. $\frac{2}{9}$ $\frac{7}{20}$ $6\frac{11}{28}$ $3\frac{13}{40}$ $4\frac{9}{32}$
      $+\frac{1}{9}$ $+\frac{9}{20}$ $+7\frac{10}{28}$ $+7\frac{17}{40}$ $+4\frac{11}{32}$

Sometimes the total in an addition problem is an improper fraction. Change the improper fraction to a mixed number. (Review page 65.) Then add the whole number part of the mixed number to the other whole numbers in the problem.

EXAMPLE: Add $4\frac{7}{8}+9\frac{5}{8}=$

$\begin{array}{r} 4\frac{7}{8} \\ +\ 9\frac{5}{8} \\ \hline 13\frac{12}{8}=14\frac{4}{8}=14\frac{1}{2} \end{array}$

Step 1.   Add the numerators. $7+5=12$

Step 2.   Write the total, 12, over the denominator.

Step 3.   Add the whole numbers. $4+9=13$

Step 4.   Change $\frac{12}{8}$ to a mixed number. $1\frac{4}{8}$

Step 5.   Add the 1 from $1\frac{4}{8}$ to 13.

Step 6.   Reduce the fraction by 4.

Add and reduce. Answers are on page 166.

7. $4\frac{3}{10}$ $9\frac{5}{6}$ $6\frac{2}{3}$ $2\frac{7}{18}$ $7\frac{7}{12}$
      $+5\frac{9}{10}$ $+3\frac{5}{6}$ $+7\frac{2}{3}$ $+8\frac{17}{18}$ $+7\frac{5}{12}$

8. $5\frac{6}{11}$ $2\frac{4}{7}$ $8\frac{3}{4}$ $3\frac{7}{9}$ $6\frac{11}{15}$
      $+3\frac{9}{11}$ $+4\frac{5}{7}$ $+6\frac{3}{4}$ $+3\frac{5}{9}$ $+9\frac{14}{15}$

9. $8\frac{4}{5}$ $3\frac{1}{2}$ $4\frac{11}{16}$ $1\frac{17}{20}$ $5\frac{19}{24}$
      $+8\frac{3}{5}$ $+7\frac{1}{2}$ $+2\frac{13}{16}$ $+9\frac{9}{20}$ $+8\frac{13}{24}$

Addition of Fractions with Different Denominators

Often the fractions in an addition problem will not have the same denominators. In these problems you must find a **common denominator.** A common denominator is a number that can be divided evenly by every denominator in the problem. The lowest number that can be divided evenly by every denominator in the problem is called the **lowest common denominator** or **LCD**. You already found common denominators when you compared fractions on pages 67 and 68.

In some problems the largest denominator in the problem is the lowest common denominator.

EXAMPLE: Add $\dfrac{5}{12} + \dfrac{1}{3} =$

$$\begin{array}{r} \dfrac{5}{12} \\ + \dfrac{1}{3} \\ \hline \end{array}$$

Step 1. Find the lowest common denomintor. 3 divides evenly into 12. 12 is the LCD.

$\dfrac{5}{12} = \dfrac{5}{12}$

Step 2. Raise $\frac{1}{3}$ to a fraction with 12 as the denominator. (Review page 64).

$+ \dfrac{1}{3} = \dfrac{4}{12}$

Step 3. Add the new fractions $\frac{5}{12}$ and $\frac{4}{12}$.

$\dfrac{9}{12} = \dfrac{3}{4}$

Step 4. Reduce the answer by 3.

Change every improper fraction answer to a mixed number and reduce. Answers are on page 166.

1.
$\begin{array}{r}\frac{2}{3}\\+\frac{5}{6}\\\hline\end{array}$
$\begin{array}{r}\frac{4}{9}\\+\frac{2}{3}\\\hline\end{array}$
$\begin{array}{r}\frac{7}{8}\\+\frac{3}{4}\\\hline\end{array}$
$\begin{array}{r}\frac{3}{10}\\+\frac{4}{5}\\\hline\end{array}$
$\begin{array}{r}\frac{5}{12}\\+\frac{3}{4}\\\hline\end{array}$

2.
$\begin{array}{r}\frac{2}{5}\\+\frac{11}{15}\\\hline\end{array}$
$\begin{array}{r}\frac{11}{18}\\+\frac{5}{9}\\\hline\end{array}$
$\begin{array}{r}\frac{1}{2}\\+\frac{9}{16}\\\hline\end{array}$
$\begin{array}{r}\frac{7}{20}\\+\frac{4}{5}\\\hline\end{array}$
$\begin{array}{r}\frac{5}{8}\\+\frac{13}{24}\\\hline\end{array}$

In addition of fractions problems the largest denominator is not always the common denominator. For example, look at this problem:

$$\begin{array}{r} \dfrac{2}{3} \\ + \dfrac{3}{4} \\ \hline \end{array}$$

4 is the largest denomintor, but 3 does not divide evenly into 4. One way to find a common denominator is to multiply the denominators in the problem together. For this example: $3 \times 4 = 12$. Both 3 and 4 divide evenly into 12.

$$\begin{array}{r} \frac{2}{3} = \frac{8}{12} \\ + \frac{3}{4} = \frac{9}{12} \\ \hline \frac{17}{12} = 1\frac{5}{12} \end{array}$$

Step 1. Multiply the denominators together.
$3 \times 4 = 12$

Step 2. Raise each fraction to a new fraction with 12 as the denominator.
$\frac{2 \times 4}{3 \times 4} = \frac{8}{12}$ and $\frac{3 \times 3}{4 \times 3} = \frac{9}{12}$

Step 3. Add the new fractions.

Step 4. Change the answer to a mixed number.

Change every improper fraction answer to a mixed number and reduce. Answers are on page 166.

3.

$\begin{array}{r} \frac{3}{4} \\ + \frac{2}{5} \\ \hline \end{array}$
$\begin{array}{r} \frac{2}{3} \\ + \frac{5}{8} \\ \hline \end{array}$
$\begin{array}{r} \frac{3}{5} \\ + \frac{5}{6} \\ \hline \end{array}$
$\begin{array}{r} \frac{1}{2} \\ + \frac{4}{7} \\ \hline \end{array}$
$\begin{array}{r} \frac{7}{9} \\ + \frac{1}{2} \\ \hline \end{array}$

4.

$\begin{array}{r} \frac{2}{3} \\ + \frac{3}{10} \\ \hline \end{array}$
$\begin{array}{r} \frac{3}{5} \\ + \frac{4}{7} \\ \hline \end{array}$
$\begin{array}{r} \frac{4}{15} \\ + \frac{1}{2} \\ \hline \end{array}$
$\begin{array}{r} \frac{5}{8} \\ + \frac{3}{5} \\ \hline \end{array}$
$\begin{array}{r} \frac{2}{9} \\ + \frac{3}{4} \\ \hline \end{array}$

Sometimes in addition problems it is not a good idea to multiply the denominators together. For example, look at this problem:

$$\begin{array}{r} \frac{5}{9} \\ + \frac{7}{12} \\ \hline \end{array}$$

Multiply the denominators together and you get $9 \times 12 = 108$. 9 and 12 both divide evenly into 108, but 108 is a large number. It is easy to make mistakes using large denominators. There are lower numbers that both 9 and 12 divide into evenly. To find the lowest common denominator, go through the multiplication table of the largest denominator until you find a number that the other denominator divides into evenly (Review page 67).

$12 \times 1 = 12.$ 9 does not divide evenly into 12.
$12 \times 2 = 24.$ 9 does not divide evenly into 24.
$12 \times 3 = 36.$ 9 does divide evenly into 36.

36 is the lowest common denominator for this problem.

$$\frac{5}{9} = \frac{20}{36}$$
$$+ \frac{7}{12} = \frac{21}{36}$$
$$\frac{41}{36} = 1\frac{5}{36}$$

Step 1. Find the lowest common denominator, 36.

Step 2. Raise each fraction to a new fraction with 36 as the denominator.

Step 3. Add the new fractions.

Step 4. Change the answer to a mixed number.

Change every improper fraction answer to a mixed number and reduce. Answers are on page 166.

5.
$$\frac{4}{9}$$
$$+ \frac{5}{6}$$

$$\frac{5}{12}$$
$$+ \frac{3}{8}$$

$$\frac{8}{15}$$
$$+ \frac{7}{10}$$

$$\frac{2}{9}$$
$$+ \frac{11}{12}$$

$$\frac{7}{16}$$
$$+ \frac{1}{6}$$

6.
$$\frac{5}{8}$$
$$+ \frac{5}{6}$$

$$\frac{1}{4}$$
$$+ \frac{9}{10}$$

$$\frac{5}{6}$$
$$+ \frac{1}{4}$$

$$\frac{4}{15}$$
$$+ \frac{9}{20}$$

$$\frac{11}{20}$$
$$+ \frac{5}{8}$$

With three fractions go through the multiplication table of the largest denominator until you find a common denominator.

7.
$$\frac{5}{12}$$
$$\frac{3}{8}$$
$$+ \frac{1}{6}$$

$$\frac{1}{2}$$
$$\frac{3}{4}$$
$$+ \frac{3}{10}$$

$$\frac{7}{20}$$
$$\frac{1}{2}$$
$$+ \frac{5}{8}$$

$$\frac{2}{9}$$
$$\frac{5}{12}$$
$$+ \frac{2}{3}$$

$$\frac{5}{6}$$
$$\frac{1}{2}$$
$$+ \frac{4}{9}$$

8.
$$\frac{1}{6}$$
$$\frac{4}{5}$$
$$+ \frac{9}{10}$$

$$\frac{3}{14}$$
$$\frac{3}{7}$$
$$+ \frac{3}{4}$$

$$\frac{1}{3}$$
$$\frac{3}{4}$$
$$+ \frac{3}{8}$$

$$\frac{1}{4}$$
$$\frac{7}{10}$$
$$+ \frac{5}{6}$$

$$\frac{5}{9}$$
$$\frac{2}{3}$$
$$+ \frac{1}{2}$$

In these problems add the whole numbers with the mixed number part of each answer.

9.
$$7\frac{5}{8}$$
$$3\frac{1}{4}$$
$$+4\frac{7}{12}$$

$$9\frac{3}{4}$$
$$6\frac{3}{8}$$
$$+2\frac{3}{10}$$

$$8\frac{4}{7}$$
$$1\frac{3}{4}$$
$$+7\frac{1}{2}$$

$$3\frac{3}{8}$$
$$4\frac{7}{10}$$
$$+6\frac{4}{5}$$

10.
$$5\frac{5}{6}$$
$$2\frac{1}{4}$$
$$+1\frac{8}{15}$$

$$7\frac{5}{12}$$
$$3\frac{2}{9}$$
$$+4\frac{3}{4}$$

$$8\frac{9}{20}$$
$$2\frac{5}{8}$$
$$+5\frac{3}{5}$$

$$9\frac{2}{3}$$
$$3\frac{1}{2}$$
$$+6\frac{4}{7}$$

Rewrite each problem with fraction under fraction and whole number under whole number. Then find the lowest common denominator, add, and reduce.

11. $4\frac{2}{3} + 7\frac{5}{9} + 5\frac{1}{4} =$ $\qquad$ $9\frac{3}{8} + 6\frac{3}{4} + 8\frac{1}{6} =$

12. $2\frac{4}{5} + 8\frac{5}{6} + 9\frac{1}{2} =$ $\qquad$ $4\frac{1}{6} + 5\frac{2}{9} + 10\frac{1}{3} =$

13. $7\frac{5}{8} + 6\frac{5}{16} + 11\frac{1}{2} =$ $\qquad$ $8\frac{2}{3} + 12\frac{1}{2} + 4\frac{7}{9} =$

14. $9\frac{2}{3} + 4\frac{7}{10} + 6\frac{1}{6} =$ $\qquad$ $6\frac{5}{8} + 8\frac{1}{2} + 7\frac{2}{5} =$

Adding Fractions Applications

In these problems you'll have a chance to apply your addition of fractions skills. Watch for words like **sum** and **total**. They usually mean to add. Other words like **combine**, **complete**, **entire**, and **altogether** sometimes mean to add.

For each problem give your answer the correct label such as inches or pounds. **Answers are on page 167.**

1. The Millers spend $\frac{2}{5}$ of their income for rent, $\frac{1}{3}$ of their income for food, and $\frac{1}{5}$ of their income for gas, electricity and telephone. Altogether these items make up what fraction of the Millers' income?

2. Andy is $71\frac{1}{4}$ inches tall. His brother Fred is $1\frac{3}{8}$ inches taller. How tall is Fred?

3. Ann works part-time at the Corner Kitchen. Monday she worked $3\frac{1}{2}$ hours. Thursday she worked $2\frac{1}{3}$ hours. Friday she worked $3\frac{3}{4}$ hours. How many hours did she work that week?

4. The distance from Tom's house to the gas station is $6\frac{7}{10}$ miles. The distance from the gas station to Tom's job is $8\frac{9}{10}$ miles. What is the distance from Tom's house to his job way of the gas station?

5. Debbie bought $3\frac{1}{2}$ pounds of fish, $2\frac{5}{8}$ pounds of cheese, and $4\frac{3}{4}$ pounds of chicken. Find the combined weight of the things Debbie bought.

6. In 1970 the town of Midvale spent 1\frac{1}{4}$ million on education. In 1980 it spent 2\frac{1}{2}$ million more on education than it spent in 1970. How much did it spend on education in 1980?

7. The porch behind Clark's house was $5\frac{1}{2}$-feet wide. Clark built a $6\frac{7}{12}$-foot wide extension on the porch. How wide was the new porch?

8. One weekend Jeff put paneling on the walls of his dining room. Friday he worked $2\frac{1}{4}$ hours. Saturday he worked $6\frac{2}{3}$ hours. Sunday he worked $3\frac{1}{2}$ hours. What was the total number of hours he worked that weekend?

9. Celeste's empty suitcase weighs $6\frac{1}{4}$ pounds. She packed the suitcase with $15\frac{9}{16}$ pounds of clothes. How much did the suitcase weigh when it was full of clothes?

10. It usually takes John $\frac{3}{4}$ of an hour to drive to work. One day because of flooding on the highway it took him $1\frac{1}{2}$ hours longer than usual. How long did it take John to drive to work that day?

11. Mrs. Gold sent three packages in the mail. One package weighed $5\frac{3}{16}$ pounds. Another weighed $2\frac{7}{16}$ pounds. The third weighed $1\frac{9}{16}$ pounds. What was the total weight of the three packages?

Subtraction of Fractions with Like Denominators

It is easy to subtract fractions with the same denominators. Subtract the numerators. Then write the difference (the answer) over the denominator.

EXAMPLE: Subtract $\dfrac{5}{6} - \dfrac{1}{6} =$

$$\begin{array}{r} \dfrac{5}{6} \\[2mm] -\dfrac{1}{6} \\[2mm] \hline \dfrac{4}{6} = \dfrac{2}{3} \end{array}$$

Step 1.   Subtract the numerators. $5 - 1 = 4$

Step 2.   Write the difference, 4, over the denominator.

Step 3.   Reduce the answer. $\frac{4 \div 2}{6 \div 2} = \frac{2}{3}$

With mixed numbers, subtract the fractions and the whole numbers separately.

EXAMPLE: Subtract $7\dfrac{8}{9} - 2\dfrac{2}{9} =$

$$\begin{array}{r} 7\dfrac{8}{9} \\[2mm] -2\dfrac{2}{9} \\[2mm] \hline 5\dfrac{6}{9} = 5\dfrac{2}{3} \end{array}$$

Step 1.   Subtract the numerators. $8 - 2 = 6$

Step 2.   Write the difference, 6, over the denominator.

Step 3.   Subtract the whole numbers. $7 - 2 = 5$

Step 4.   Reduce the answer. $5\,\frac{6 \div 3}{9 \div 3} = 5\frac{2}{3}$

Subtract and reduce each problem. Answers are on page 167.

1.
$\begin{array}{r} \frac{11}{15} \\[1mm] -\frac{2}{15} \\ \hline \end{array}$
$\qquad$
$\begin{array}{r} \frac{7}{10} \\[1mm] -\frac{3}{10} \\ \hline \end{array}$
$\qquad$
$\begin{array}{r} 9\frac{8}{9} \\[1mm] -\ 2\frac{5}{9} \\ \hline \end{array}$
$\qquad$
$\begin{array}{r} 8\frac{9}{14} \\[1mm] -3\frac{3}{14} \\ \hline \end{array}$
$\qquad$
$\begin{array}{r} 6\frac{17}{21} \\[1mm] -\ 5\frac{11}{21} \\ \hline \end{array}$

2.
$\begin{array}{r} \frac{7}{8} \\[1mm] -\frac{1}{8} \\ \hline \end{array}$
$\qquad$
$\begin{array}{r} \frac{15}{16} \\[1mm] -\frac{7}{16} \\ \hline \end{array}$
$\qquad$
$\begin{array}{r} 12\frac{3}{4} \\[1mm] -\ 3\frac{1}{4} \\ \hline \end{array}$
$\qquad$
$\begin{array}{r} 7\frac{9}{20} \\[1mm] -4\frac{3}{20} \\ \hline \end{array}$
$\qquad$
$\begin{array}{r} 11\frac{19}{25} \\[1mm] -\ 9\frac{4}{25} \\ \hline \end{array}$

3.
$\begin{array}{r} \frac{11}{12} \\[1mm] -\frac{7}{12} \\ \hline \end{array}$
$\qquad$
$\begin{array}{r} \frac{5}{6} \\[1mm] -\frac{1}{6} \\ \hline \end{array}$
$\qquad$
$\begin{array}{r} 5\frac{11}{18} \\[1mm] -\ 1\frac{5}{18} \\ \hline \end{array}$
$\qquad$
$\begin{array}{r} 6\frac{23}{30} \\[1mm] -3\frac{11}{30} \\ \hline \end{array}$
$\qquad$
$\begin{array}{r} 13\frac{13}{24} \\[1mm] -\ 8\frac{5}{24} \\ \hline \end{array}$

Subtraction of Fractions with Unlike Denominators

Often the fractions in a subtraction problem will not have the same denominators. In these problems find a common denominator. Change each fraction to a new fraction with the common denominator. Then subtract.

EXAMPLE: Subtract $6\frac{3}{4} - 2\frac{2}{5} =$

$$6\frac{3}{4}$$
$$-2\frac{2}{5}$$

Step 1.   Find the common denominator. 4 and 5 both divide evenly into 20. 20 is the common denominator.

$$6\frac{3}{4} = 6\frac{15}{20}$$
$$-2\frac{2}{5} = 2\frac{8}{20}$$

Step 2.   Raise $\frac{3}{4}$ and $\frac{2}{5}$ to new fractions with 20 as the denominator. $\frac{3 \times 5}{4 \times 5} = \frac{15}{20}$ and $\frac{2 \times 4}{5 \times 4} = \frac{8}{20}$

Step 3.   Subtract the new fractions.

$$4\frac{7}{20}$$

Step 4.   Subtract the whole numbers.

Subtract and reduce each problem. Answers are on page 167.

1.
$$\frac{7}{8}$$
$$-\frac{2}{3}$$

$$\frac{3}{5}$$
$$-\frac{3}{10}$$

$$8\frac{2}{3}$$
$$-1\frac{1}{4}$$

$$7\frac{5}{9}$$
$$-3\frac{1}{3}$$

$$9\frac{1}{2}$$
$$-5\frac{1}{6}$$

2.
$$\frac{1}{2}$$
$$-\frac{2}{5}$$

$$\frac{6}{7}$$
$$-\frac{1}{3}$$

$$9\frac{3}{4}$$
$$-6\frac{3}{8}$$

$$6\frac{4}{5}$$
$$-4\frac{2}{3}$$

$$10\frac{3}{4}$$
$$-1\frac{2}{9}$$

3.
$$\frac{5}{6}$$
$$-\frac{1}{4}$$

$$\frac{7}{9}$$
$$-\frac{1}{2}$$

$$4\frac{1}{2}$$
$$-3\frac{1}{4}$$

$$11\frac{7}{12}$$
$$-9\frac{3}{8}$$

$$3\frac{5}{6}$$
$$-1\frac{3}{10}$$

Borrowing

Sometimes there is no top fraction to subtract the bottom fraction from. To get a fraction in the top number, you have to **borrow.** To borrow means to take 1 from the whole number at the top. Then rewrite 1 as a fraction. Remember that any fraction with the same numerator and denominator is equal to 1. For example, $\frac{6}{6} = 1$; $\frac{5}{5} = 1$; $\frac{4}{4} = 1$; and so on. The numerator and denominator should be the same as the denominator of the other fraction in the problem.

EXAMPLE: Subtract $8 - 2\frac{5}{9} =$

$$8$$
$$-2\frac{5}{9}$$

Step 1. Borrow 1 from the 8. Change the 1 to $\frac{9}{9}$ since 9 is the common denominator.

Step 2. Subtract the fractions. $\frac{9}{9} - \frac{5}{9} = \frac{4}{9}$

$$\overset{7}{\cancel{8}}\frac{9}{9}$$
$$-2\frac{5}{9}$$
$$\overline{5\frac{4}{9}}$$

Step 3. Subtract the whole numbers. $7 - 2 = 5$

Borrowing often causes problems. Be sure you understand the last example before you try these problems.

Subtract and reduce each problem. Answers are on page 167.

1.
$$\begin{array}{r} 6 \\ -2\frac{3}{7} \\ \hline \end{array}$$
$$\begin{array}{r} 9 \\ -7\frac{7}{8} \\ \hline \end{array}$$
$$\begin{array}{r} 5 \\ -1\frac{7}{9} \\ \hline \end{array}$$
$$\begin{array}{r} 4 \\ -1\frac{3}{4} \\ \hline \end{array}$$
$$\begin{array}{r} 8 \\ -3\frac{1}{2} \\ \hline \end{array}$$

2.
$$\begin{array}{r} 7 \\ -5\frac{7}{10} \\ \hline \end{array}$$
$$\begin{array}{r} 9 \\ -2\frac{5}{12} \\ \hline \end{array}$$
$$\begin{array}{r} 10 \\ -8\frac{9}{16} \\ \hline \end{array}$$
$$\begin{array}{r} 3 \\ -1\frac{8}{15} \\ \hline \end{array}$$
$$\begin{array}{r} 14 \\ -6\frac{11}{20} \\ \hline \end{array}$$

Sometimes the top fraction is not big enough to subtract the bottom fraction from. To get a bigger fraction in the top position, borrow 1 from the whole number at the top. Rewrite the 1 as a fraction. Then add this new fraction to the old top fraction. Study this example carefully.

EXAMPLE: Subtract $5\frac{1}{8} - 2\frac{7}{8} =$

$$5\frac{1}{8}$$
$$-2\frac{7}{8}$$

Step 1. Borrow 1 from the 5. Rewrite the 1 as $\frac{8}{8}$ since 8 is the common denominator.

$$\overset{4}{\cancel{5}}\frac{1}{8} + \frac{8}{8} = 4\frac{9}{8}$$
$$-2\frac{7}{8} \qquad 2\frac{7}{8}$$

Step 2. Add $\frac{8}{8}$ to $\frac{1}{8}$. $\frac{8}{8} + \frac{1}{8} = \frac{9}{8}$

Step 3. Subtract the new fractions.

Step 4. Subtract the new whole numbers.

$$2\frac{2}{8} = 2\frac{1}{4}$$

Step 5. Reduce the answer.

Be sure you understand this example before you try these problems.

Subtract and reduce each problem. Answers are on page 167.

3.
$$9\frac{3}{7} \qquad 5\frac{4}{9} \qquad 8\frac{2}{5} \qquad 7\frac{1}{4} \qquad 4\frac{3}{8}$$
$$-4\frac{6}{7} \qquad -1\frac{7}{9} \qquad -3\frac{4}{5} \qquad -2\frac{3}{4} \qquad -2\frac{7}{8}$$

4.
$$6\frac{5}{16} \qquad 10\frac{1}{6} \qquad 9\frac{3}{10} \qquad 12\frac{1}{3} \qquad 8\frac{5}{12}$$
$$-3\frac{9}{16} \qquad -2\frac{5}{6} \qquad -6\frac{9}{10} \qquad -3\frac{2}{3} \qquad -5\frac{11}{12}$$

When the denominators in a subtraction problem are not the same, change each fraction to a new fraction with a common denominator. Then borrow if you need to.

EXAMPLE: Subtract $8\frac{1}{3} - 4\frac{7}{9} =$

$$8\frac{1}{3}$$
$$-4\frac{7}{9}$$

Step 1. Find the lowest common denominator. 3 divides evenly into 9. 9 is the LCD.

$$8\frac{1}{3} = 8\frac{3}{9}$$
$$-4\frac{7}{9} = 4\frac{7}{9}$$

Step 2. Raise $\frac{1}{3}$ to a fraction with 9 as the denominator. $\frac{1 \times 3}{3 \times 3} = \frac{3}{9}$

Step 3. Borrow 1 from the 8. Change the 1 to $\frac{9}{9}$ since 9 is the common denominator.

$$\overset{7}{\cancel{8}}\frac{3}{9} + \frac{9}{9} = 7\frac{12}{9}$$
$$-4\frac{7}{9} \qquad\quad = 4\frac{7}{9}$$
$$\overline{}$$
$$3\frac{5}{9}$$

Step 4. Add $\frac{9}{9}$ to $\frac{3}{9}$. $\frac{3}{9} + \frac{9}{9} = \frac{12}{9}$

Step 5. Subtract the new fractions.

Step 6. Subtract the whole numbers.

Step 7. Reduce the answer. $3\frac{5}{9}$ is reduced.

Subtract and reduce each problem. Answers are on page 167.

5. $8\frac{1}{2}$ $9\frac{3}{5}$ $7\frac{1}{3}$ $6\frac{1}{2}$ $5\frac{2}{9}$

     $-5\frac{2}{3}$ $-2\frac{3}{4}$ $-3\frac{5}{8}$ $-4\frac{7}{8}$ $-1\frac{5}{6}$

6. $3\frac{1}{4}$ $8\frac{3}{7}$ $9\frac{5}{12}$ $10\frac{1}{3}$ $7\frac{2}{7}$

     $-1\frac{5}{6}$ $-6\frac{2}{3}$ $-4\frac{3}{4}$ $-3\frac{4}{7}$ $-2\frac{3}{4}$

7. $6\frac{1}{3}$ $12\frac{3}{8}$ $4\frac{1}{6}$ $9\frac{1}{2}$ $13\frac{1}{3}$

     $-1\frac{8}{15}$ $-8\frac{7}{10}$ $-1\frac{4}{5}$ $-7\frac{11}{18}$ $-6\frac{5}{9}$

Rewrite each problem with fraction under fraction and whole number under whole number. Then find the lowest common denominator, subtract, and reduce. Answers are on page 167.

8. $7\frac{2}{5} - 2\frac{5}{6} =$ $6\frac{3}{4} - 3\frac{2}{5} =$ $9\frac{1}{2} - 1\frac{1}{3} =$

9. $12\frac{1}{2} - 4\frac{7}{9} =$ $8\frac{1}{3} - 5\frac{7}{12} =$ $4\frac{1}{2} - 2\frac{14}{15} =$

10. $9\frac{6}{7} - 6\frac{5}{21} =$ $7\frac{1}{2} - 1\frac{3}{4} =$ $12\frac{4}{5} - 3\frac{3}{10} =$

11. $8\frac{3}{4} - 2\frac{19}{40} =$ $6\frac{7}{20} - 4\frac{3}{50} =$ $10\frac{1}{3} - 5\frac{11}{15} =$

Subtracting Fractions Applications

These problems give you a chance to apply your subtraction of fractions skills. Watch for words like **difference** and **balance**. They usually mean to subtract. Phrases like **how much greater** and **how much less** also mean to subtract.

Give your answers the correct units such as feet or ounces. **Answers are on page 167.**

1. The Midvale City Council needs $4 million to build a new hospital. So far it has collected $$1\frac{7}{8}$ million. Find the balance that it needs to raise.

2. Mike is $70\frac{1}{2}$ inches tall. His wife, Maureen, is $67\frac{3}{4}$ inches tall. How much taller is Mike than Maureen?

3. Jose bought a piece of lumber $95\frac{1}{2}$ inches long. He sawed off a piece that was $37\frac{5}{8}$ inches long. How long was the piece that was left?

4. Valerie used $3\frac{1}{4}$ pounds of flour to bake cakes for her children's school fair. She took the flour from a five-pound bag. How much flour did she have left?

5. The distance from Mark's house to his factory is $8\frac{3}{10}$ miles. On his way to work Mark stops to pick up Al. Al lives $3\frac{9}{10}$ miles from Mark's house. How far is it from Al's house to the factory?

6. In 1970 the population of Central County was $1\frac{5}{6}$ million people. In 1980 the population was $2\frac{1}{3}$ million. How much greater was the population of the county in 1980?

7. Marvin weighed 243 pounds. He went on a diet and lost $54\frac{1}{2}$ pounds. How much did Marvin weigh after his diet?

8. The operating budget for the town of Midvale was $\$4\frac{1}{5}$ million for 1980. By June of 1980 the town had spent $\$2\frac{4}{5}$ million of the budget. Find the balance of the budget that was left for the rest of the year.

9. Patty usually works 35 hours a week. Last week she was sick and missed $12\frac{1}{4}$ hours of work. How many hours did Patty work last week?

10. Dorothy and Robert spend $\frac{1}{4}$ of their income for mortgage payments. They try to save $\frac{1}{10}$ of their income. What is the difference between the fraction they spend on the mortgage and the fraction they save?

11. Sometimes Bill drives to work. Driving usually takes him $1\frac{1}{4}$ hours. Other times he take the train. The train takes him $1\frac{5}{12}$ hours. How much faster is driving than taking the train?

Multiplication of Fractions

When you multiply whole numbers (except 1 and 0), the answer is bigger than the two numbers you multiply. The opposite is true when you multiply fractions. When you multiply two fractions, the answer is smaller than either of the two fractions you multiplied. When you multiply two fractions, you are finding **a part of a part**. For example, if you multiply $\frac{1}{2}$ by $\frac{1}{2}$, you are finding $\frac{1}{2}$ **of** $\frac{1}{2}$. You know that $\frac{1}{2}$ of $\frac{1}{2}$ dollar is $\frac{1}{4}$ dollar. The answer is smaller than either of the fractions you multiplied.

Multiplying fractions is easier than adding or subtracting fractions. To multiply fractions you do not first need a common denominator. To multiply fractions, multiply the numerators together and multiply the denominators together. Then reduce the answer.

EXAMPLE: Multiply $\dfrac{5}{8} \times \dfrac{7}{9} =$

$\dfrac{5}{8} \times \dfrac{7}{9} = \dfrac{35}{72}$

Step 1.   Multiply the numerators. $5 \times 7 = 35$

Step 2.   Multiply the denominators. $8 \times 9 = 72$

Step 3.   Try to reduce. $\frac{35}{72}$ is reduced.

Multiply and reduce. Answers are on page 167.

1. $\dfrac{1}{5} \times \dfrac{2}{3} =$ $\qquad$ $\dfrac{3}{8} \times \dfrac{3}{5} =$ $\qquad$ $\dfrac{4}{5} \times \dfrac{3}{7} =$ $\qquad$ $\dfrac{1}{3} \times \dfrac{4}{5} =$

2. $\dfrac{5}{6} \times \dfrac{1}{8} =$ $\qquad$ $\dfrac{3}{4} \times \dfrac{7}{10} =$ $\qquad$ $\dfrac{4}{9} \times \dfrac{2}{5} =$ $\qquad$ $\dfrac{1}{7} \times \dfrac{1}{8} =$

3. $\dfrac{7}{9} \times \dfrac{5}{8} =$ $\qquad$ $\dfrac{3}{4} \times \dfrac{3}{7} =$ $\qquad$ $\dfrac{1}{3} \times \dfrac{8}{9} =$ $\qquad$ $\dfrac{5}{8} \times \dfrac{1}{3} =$

4. $\dfrac{2}{5} \times \dfrac{4}{5} =$ $\qquad$ $\dfrac{8}{15} \times \dfrac{2}{5} =$ $\qquad$ $\dfrac{6}{7} \times \dfrac{5}{11} =$ $\qquad$ $\dfrac{5}{8} \times \dfrac{5}{9} =$

Canceling

Canceling is a way of making many multiplication of fractions problems easier. It is like reducing. To cancel, divide a numerator and a denominator by a number that goes evenly into both to them.

EXAMPLE: Multiply $\frac{4}{9} \times \frac{6}{7} =$

$$\frac{4}{\overset{}{\underset{3}{9}}} \times \frac{\overset{2}{6}}{7} = \frac{8}{21}$$

Step 1. Cancel: Divide 3 into 6 and 9. Cross out 6 and write 2. Cross out 9 and write 3.

Step 2. Multiply the new numerators. $4 \times 2 = 8$

Step 3. Multiply the new denominators. $3 \times 7 = 21$

Step 4. Try to reduce. $\frac{8}{21}$ is reduced.

Canceling is not necessary, but it makes problems easier. You could multiply the last example without canceling:

$$\frac{4}{9} \times \frac{6}{7} = \frac{24}{63} \quad \text{Reduce} \quad \frac{24}{63} \text{ by 3:} \quad \frac{24 \div 3}{63 \div 3} = \frac{8}{21}$$

Sometimes it is possible to cancel more than once.

EXAMPLE: Multiply $\frac{8}{9} \times \frac{3}{20} =$

$$\frac{\overset{2}{8}}{\underset{3}{9}} \times \frac{\overset{1}{3}}{\underset{5}{20}} = \frac{2}{15}$$

Step 1. Cancel: Divide 4 into 8 and 20. Cross out 8 and write 2. Cross out 20 and write 5.

Step 2. Cancel: Divide 3 into 3 and 9. Cross out 3 and write 1. Cross out 9 and write 3.

Step 3. Multiply the new numerators.

Step 4. Multiply the new denominators.

Step 5. Try to reduce. $\frac{2}{15}$ is reduced.

In the last example, canceling made the problem much easier. Without canceling:

$$\frac{8}{9} \times \frac{3}{20} = \frac{24}{180} \quad \text{Reduce} \quad \frac{24}{180} \text{ by 12:} \quad \frac{24 \div 12}{180 \div 12} = \frac{2}{15}$$

Cancel, multiply, and reduce. Answers are on page 168.

1. $\dfrac{4}{9} \times \dfrac{6}{7} =$ $\dfrac{5}{6} \times \dfrac{3}{7} =$ $\dfrac{2}{3} \times \dfrac{7}{8} =$ $\dfrac{3}{4} \times \dfrac{8}{15} =$

2. $\dfrac{5}{9} \times \dfrac{3}{20} =$ $\dfrac{4}{9} \times \dfrac{3}{8} =$ $\dfrac{7}{10} \times \dfrac{8}{21} =$ $\dfrac{5}{12} \times \dfrac{6}{13} =$

3. $\dfrac{5}{18} \times \dfrac{4}{15} =$ $\dfrac{8}{9} \times \dfrac{9}{10} =$ $\dfrac{8}{15} \times \dfrac{9}{14} =$ $\dfrac{4}{21} \times \dfrac{9}{16} =$

4. $\dfrac{4}{9} \times \dfrac{5}{14} =$ $\dfrac{8}{33} \times \dfrac{11}{24} =$ $\dfrac{9}{10} \times \dfrac{4}{9} =$ $\dfrac{6}{7} \times \dfrac{3}{8} =$

5. $\dfrac{9}{20} \times \dfrac{8}{15} =$ $\dfrac{7}{20} \times \dfrac{5}{14} =$ $\dfrac{5}{7} \times \dfrac{21}{25} =$ $\dfrac{2}{15} \times \dfrac{3}{8} =$

6. $\dfrac{4}{5} \times \dfrac{15}{16} =$ $\dfrac{9}{40} \times \dfrac{2}{3} =$ $\dfrac{1}{2} \times \dfrac{8}{9} =$ $\dfrac{15}{16} \times \dfrac{4}{5} =$

7. $\dfrac{24}{25} \times \dfrac{5}{8} =$ $\dfrac{7}{9} \times \dfrac{12}{35} =$ $\dfrac{10}{11} \times \dfrac{3}{10} =$ $\dfrac{11}{30} \times \dfrac{5}{33} =$

8. $\dfrac{15}{16} \times \dfrac{4}{9} =$ $\dfrac{3}{8} \times \dfrac{8}{9} =$ $\dfrac{3}{16} \times \dfrac{5}{6} =$ $\dfrac{9}{10} \times \dfrac{2}{3} =$

Multiplication with Fractions and Whole Numbers

To multiply a whole number and a fraction, first write the whole number as a fraction. Write the whole number as the numerator and 1 as the denominator.

EXAMPLE: Multiply $\dfrac{5}{6} \times 9 =$

$$\dfrac{5}{\cancel{6}_{2}} \times \dfrac{\cancel{9}^{3}}{1} = \dfrac{15}{2} = 7\dfrac{1}{2}$$

Step 1. Write 9 as a fraction with a denominator of 1. $\dfrac{9}{1}$

Step 2. Cancel: Divide 9 and 6 by 3. Cross out 9 and write 3. Cross out 6 and write 2.

Step 3. Multiply the new numerators. $5 \times 3 = 15$

Step 4. Multiply the new denominators. $2 \times 1 = 2$

Step 5. Change the answer, $\dfrac{15}{2}$, to a mixed number, $7\dfrac{1}{2}$. (Review page 65.)

Cancel, multiply, and reduce. Change every improper fraction answer to a mixed number. Answers are on page 168.

1. $8 \times \dfrac{3}{4} =$ $\qquad$ $15 \times \dfrac{2}{5} =$ $\qquad$ $\dfrac{5}{9} \times 12 =$ $\qquad$ $\dfrac{1}{6} \times 9 =$

2. $\dfrac{4}{5} \times 10 =$ $\qquad$ $\dfrac{5}{12} \times 20 =$ $\qquad$ $6 \times \dfrac{2}{3} =$ $\qquad$ $5 \times \dfrac{3}{10} =$

3. $2 \times \dfrac{7}{8} =$ $\qquad$ $7 \times \dfrac{5}{14} =$ $\qquad$ $\dfrac{3}{8} \times 16 =$ $\qquad$ $\dfrac{8}{15} \times 18 =$

4. $\dfrac{5}{16} \times 24 =$ $\qquad$ $\dfrac{7}{12} \times 30 =$ $\qquad$ $3 \times \dfrac{2}{9} =$ $\qquad$ $4 \times \dfrac{7}{8} =$

Multiplication with Mixed Numbers

To multiply mixed numbers, first change the mixed numbers to improper fractions. (Review page 66.) Then multiply the improper fractions.

EXAMPLE: Multiply $\frac{3}{7} \times 1\frac{5}{9} =$

$\frac{3}{7} \times 1\frac{5}{9} =$ *Step 1.*   Change $1\frac{5}{9}$ to a fraction: $9 \times 1 = 9$
                              $9 + 5 = 14. \frac{14}{9}$

$\frac{3}{7} \times \frac{14}{9} =$ *Step 2.*   Cancel: Divide 3 and 9 by 3. Cross out 3 and write 1. Cross out 9 and write 3.

$\overset{1}{\underset{1}{\cancel{\frac{3}{7}}}} \times \overset{2}{\underset{3}{\cancel{\frac{14}{9}}}} = \frac{2}{3}$ *Step 3.*   Cancel: Divide 14 and 7 by 7. Cross out 14 and write 2. Cross out 7 and write 1.

                       Step 4.   Multiply the new numerators.

                       Step 5.   Multiply the new denominators.

                       Step 6.   Try to reduce. $\frac{2}{3}$ is reduced.

Change each mixed number to an improper fraction, cancel, multiply, and reduce. Change each improper fraction answer to a mixed number. Answers are on page 168.

1. $\frac{8}{9} \times 3\frac{3}{4} =$ $2\frac{1}{12} \times \frac{9}{10} =$ $6\frac{2}{3} \times \frac{15}{16} =$ $\frac{4}{9} \times 2\frac{5}{8} =$

2. $4\frac{2}{3} \times 6 =$ $12 \times 1\frac{3}{4} =$ $1\frac{3}{10} \times 5 =$ $9 \times 2\frac{1}{6} =$

3. $\frac{2}{3} \times 4\frac{4}{5} =$ $2\frac{4}{7} \times \frac{7}{9} =$ $\frac{3}{4} \times 2\frac{2}{9} =$ $3\frac{1}{5} \times \frac{5}{6} =$

4. $8 \times 2\frac{1}{4} =$ $\qquad$ $5 \times 2\frac{7}{10} =$ $\qquad$ $1\frac{2}{3} \times 9 =$ $\qquad$ $5\frac{3}{8} \times 4 =$

5. $4\frac{2}{3} \times \frac{6}{7} =$ $\qquad$ $\frac{3}{8} \times 3\frac{1}{5} =$ $\qquad$ $6\frac{2}{3} \times \frac{3}{4} =$ $\qquad$ $\frac{9}{20} \times 5\frac{1}{3} =$

6. $\frac{5}{6} \times 1\frac{5}{7} =$ $\qquad$ $1\frac{7}{8} \times \frac{4}{9} =$ $\qquad$ $\frac{5}{7} \times 1\frac{3}{10} =$ $\qquad$ $2\frac{2}{5} \times \frac{7}{8} =$

7. $1\frac{7}{15} \times 1\frac{1}{8} =$ $\qquad$ $3\frac{3}{5} \times 3\frac{3}{4} =$ $\qquad$ $5\frac{1}{4} \times 1\frac{13}{15} =$ $\qquad$ $1\frac{7}{8} \times 3\frac{3}{5} =$

8. $1\frac{1}{5} \times 1\frac{1}{3} =$ $\qquad$ $2\frac{4}{5} \times 1\frac{3}{7} =$ $\qquad$ $1\frac{2}{7} \times 5\frac{1}{4} =$ $\qquad$ $1\frac{3}{11} \times 1\frac{5}{6} =$

9. $2\frac{2}{5} \times 3\frac{3}{4} =$ $\qquad$ $2\frac{2}{9} \times 2\frac{5}{8} =$ $\qquad$ $1\frac{1}{7} \times 1\frac{5}{16} =$ $\qquad$ $7\frac{1}{2} \times 1\frac{1}{3} =$

10. $1\frac{7}{8} \times 2\frac{2}{3} =$ $\qquad$ $2\frac{1}{4} \times 4\frac{2}{3} =$ $\qquad$ $13\frac{1}{3} \times 2\frac{2}{5} =$ $\qquad$ $1\frac{3}{4} \times 1\frac{5}{7} =$

Multiplying Fractions Applications

These problems give you a chance to apply your multiplication of fractions skills. The key word that tells you to multiply is **of**. A fraction immediately followed by the word **of** means to multiply. Also, watch for situations that mean to multiply. For example, you may be given the weight of one thing or the cost of one thing. Then you may be asked to find the weight or cost of several things. These problems mean to multiply.

Give every answer the correct label such as dollars or ounces. **Answers are on page 168.**

1. Joe's weekly salary is $282. His employer takes out $\frac{1}{6}$ of Joe's salary for taxes and Social Security. How much does Joe's employer take out of Joe's weekly check?

2. One can of tomato sauce weighs $4\frac{3}{4}$ ounces. What is the weight of the cans in a box that holds 16 cans of tomato sauce?

3. Pete bought $10\frac{1}{2}$ feet of lumber. The lumber cost $2.40 a foot. How much did Pete pay for all the lumber he bought?

4. Sarah had 5 pounds of sugar. She used $\frac{3}{8}$ of the sugar for baking. How much sugar did Sarah use?

5. One pound of tomatoes costs 28¢. How much do $2\frac{3}{4}$ pounds of tomatoes cost?

6. In June 560 people took driving tests in the town of Midvale. $\frac{7}{10}$ of the people passed the test. How many people passed the driving test in June?

7. The Oakdale Block Association wants to raise $4,000 to improve the playground on its street. After a month of working, it has collected $\frac{5}{8}$ of the money. How much money does it have now?

8. Nate runs a used car lot. On July 1 he had 64 used cars to sell. By July 30 he had sold $\frac{3}{4}$ of them. How many cars did he sell in July?

9. Susan makes $5.80 an hour for overtime work. Last week she worked $6\frac{1}{2}$ hours overtime. How much did she make for overtime work last week?

10. Gordon makes furniture. He needs $7\frac{1}{3}$ feet of lumber to build a cabinet. How much lumber does he need to build four cabinets?

11. George and Margie want to save $6,500 to make a down payment on a house. So far they have saved $\frac{4}{5}$ of what they need. How much have they saved?

12. Alice bought $5\frac{1}{4}$ yards of material. The material cost $6.80 a yard. How much did she pay for the material?

Division by Fractions

Dividing by a fraction means finding how many times a fraction goes into another number. For example, if you divide $\frac{1}{2}$ by $\frac{1}{4}$, you find out how many times $\frac{1}{4}$ goes into $\frac{1}{2}$. You know that $\frac{1}{4}$ dollar goes into $\frac{1}{2}$ dollar exactly twice. We write this problem as $\frac{1}{2} \div \frac{1}{4}$. Read the $\div$ sign with the words "divided by." The fraction at the right of the $\div$ sign is the **divisor**.

To divide fractions, **invert** the divisor and follow the rules for multiplying fractions. To invert means to turn a fraction around. Write the numerator on the bottom and the denominator on the top. For example, when you invert $\frac{2}{3}$, you get $\frac{3}{2}$.

EXAMPLE: Divide $\frac{5}{9} \div \frac{2}{3} =$

$\frac{5}{9} \div \frac{2}{3} =$

$\frac{5}{\cancel{9}_3} \times \frac{\cancel{3}^1}{2} = \frac{5}{6}$

Step 1. Invert the divisor $\frac{2}{3}$ to $\frac{3}{2}$ and change the division sign ($\div$) to a multiplication sign ($\times$).

Step 2. Cancel: Divide 3 and 9 by 3. Cross out 3 and write 1. Cross out 9 and write 3.

Step 3. Multiply the new numerators. $5 \times 1 = 5$

Step 4. Multiply the new denominators. $3 \times 2 = 6$

Step 5. Try to reduce. $\frac{5}{6}$ is reduced.

Change each improper fraction answer to a whole or mixed number. Answers are on page 168.

1. $\frac{9}{10} \div \frac{3}{4} =$ $\frac{1}{2} \div \frac{3}{8} =$ $\frac{6}{7} \div \frac{3}{5} =$ $\frac{5}{8} \div \frac{1}{12} =$

2. $\frac{4}{9} \div \frac{2}{3} =$ $\frac{1}{4} \div \frac{5}{12} =$ $\frac{5}{8} \div \frac{15}{16} =$ $\frac{3}{10} \div \frac{2}{5} =$

3. $\frac{9}{20} \div \frac{3}{4} =$ $\frac{5}{6} \div \frac{4}{9} =$ $\frac{3}{5} \div \frac{6}{7} =$ $\frac{9}{16} \div \frac{3}{20} =$

When you divide a whole number by a fraction, you find out how many times the fraction goes into the whole number. To divide a whole number by a fraction, first write the whole number as a fraction. Write the whole number as the numerator and 1 as the denominator. Then invert the fraction at the right and follow the rules for multiplying fractions.

EXAMPLE: Divide $4 \div \frac{6}{7} =$

$4 \div \frac{6}{7} =$

Step 1. Write 4 as a fraction with a denominator of 1. $\frac{4}{1}$

$\frac{4}{1} \div \frac{6}{7} =$

Step 2. Invert the divisor $\frac{6}{7}$ to $\frac{7}{6}$ and change the division sign ($\div$) to a multiplication sign ($\times$).

$\frac{\overset{2}{\cancel{4}}}{1} \times \frac{7}{\underset{3}{\cancel{6}}} = \frac{14}{3} = 4\frac{2}{3}$

Step 3. Cancel: Divide 6 and 4 by 2. Cross out 6 and write 3. Cross out 4 and write 2.

Step 4. Multiply the new numerators. $2 \times 7 = 14$

Step 5. Multiply the new denominators. $1 \times 3 = 3$

Step 6. Change the improper fraction to a mixed number. $\frac{14}{3} = 4\frac{2}{3}$

Change every improper fraction answer to a whole or mixed number and reduce. Answers are on page 168.

4. $6 \div \frac{3}{8} =$ $\qquad$ $8 \div \frac{4}{5} =$ $\qquad$ $9 \div \frac{6}{7} =$ $\qquad$ $5 \div \frac{2}{3} =$

5. $4 \div \frac{1}{2} =$ $\qquad$ $3 \div \frac{9}{10} =$ $\qquad$ $2 \div \frac{1}{4} =$ $\qquad$ $7 \div \frac{7}{10} =$

6. $10 \div \frac{5}{6} =$ $\qquad$ $12 \div \frac{9}{20} =$ $\qquad$ $15 \div \frac{10}{11} =$ $\qquad$ $16 \div \frac{6}{7} =$

To divide a mixed number, first change the mixed number to an improper fraction. Then follow the rules for dividing by fractions.

EXAMPLE: Divide $3\frac{2}{3} \div \frac{5}{6} =$

$3\frac{2}{3} \div \frac{5}{6} =$

Step 1. Change $3\frac{2}{3}$ to an improper fraction. $3\frac{2}{3} = \frac{11}{3}$

$\frac{11}{3} \div \frac{5}{6} =$

Step 2. Invert the divisor $\frac{5}{6}$ to $\frac{6}{5}$ and change the division sign ($\div$) to a multiplication sign ($\times$).

$\frac{11}{\cancel{3}_1} \times \frac{\cancel{6}^2}{5} = \frac{22}{5} = 4\frac{2}{5}$

Step 3. Cancel: Divide 6 and 3 by 3. Cross out 6 and write 2. Cross out 3 and write 1.

Step 4. Multiply the new numerators. $11 \times 2 = 22$

Step 5. Multiply the new denominators. $1 \times 5 = 5$

Step 6. Change the answer $\frac{22}{5}$ to a mixed number. $\frac{22}{5} = 4\frac{2}{5}$

Change every improper fraction answer to a whole or mixed number and reduce. Answers are on page 169.

7. $3\frac{1}{2} \div \frac{3}{8} =$ $4\frac{4}{5} \div \frac{2}{5} =$ $7\frac{1}{2} \div \frac{9}{16} =$ $1\frac{3}{4} \div \frac{1}{2} =$

8. $1\frac{1}{3} \div \frac{5}{9} =$ $2\frac{2}{9} \div \frac{5}{12} =$ $3\frac{1}{2} \div \frac{7}{10} =$ $2\frac{2}{9} \div \frac{2}{3} =$

9. $5\frac{1}{4} \div \frac{7}{8} =$ $3\frac{3}{4} \div \frac{5}{6} =$ $4\frac{1}{6} \div \frac{5}{12} =$ $2\frac{2}{3} \div \frac{20}{21} =$

Division of Fractions and Mixed Numbers by Whole Numbers

When you divide a fraction or a mixed number by a whole number, you "split" the fraction or mixed number into smaller parts. For example, if you divide a fraction by 2, you split the fraction into two equal parts. To divide a fraction by a whole number, first write the whole number as a fraction. Write the whole number as the numerator and 1 as the denominator. Then invert the fraction at the right and follow the rules for multiplying fractions.

EXAMPLE: Divide $\frac{7}{8} \div 2 =$

$\frac{7}{8} \div 2 =$ *Step 1.* Write 2 as a fraction with a denominator of 1. $\frac{2}{1}$

$\frac{7}{8} \div \frac{2}{1} =$ *Step 2.* Invert the divisor $\frac{2}{1}$ to $\frac{1}{2}$ and change the division sign (÷) to a multiplication sign (×).

$\frac{7}{8} \times \frac{1}{2} = \frac{7}{16}$ *Step 3.* Multiply the numerators. $7 \times 1 = 7$

Step 4. Multiply the denominators. $8 \times 2 = 16$

Step 5. Try to reduce. $\frac{7}{16}$ is reduced.

Change every improper fraction answer to a whole or mixed number and reduce. Answers are on page 169.

1. $\frac{3}{5} \div 6 =$ $\frac{4}{9} \div 10 =$ $\frac{7}{8} \div 14 =$ $\frac{1}{4} \div 3 =$

2. $\frac{4}{7} \div 12 =$ $\frac{9}{10} \div 15 =$ $\frac{5}{6} \div 4 =$ $\frac{8}{15} \div 6 =$

3. $1\frac{1}{5} \div 4 =$ $2\frac{1}{2} \div 5 =$ $1\frac{2}{3} \div 10 =$ $2\frac{2}{5} \div 2 =$

4. $1\frac{5}{7} \div 8 =$ $2\frac{1}{4} \div 3 =$ $4\frac{2}{3} \div 7 =$ $5\frac{1}{4} \div 9 =$

Division by Mixed Numbers

To divide by a mixed number, first change any mixed number in the problem to an improper fraction. Then invert the fraction on the right and follow the rules for multiplying fractions.

EXAMPLE: Divide $\frac{15}{16} \div 2\frac{1}{4} =$

$\frac{15}{16} \div 2\frac{1}{4} =$

Step 1. Change the mixed number to an improper fraction. $2\frac{1}{4} = \frac{9}{4}$

$\frac{15}{16} \div \frac{9}{4} =$

Step 2. Invert the divisor $\frac{9}{4}$ to $\frac{4}{9}$ and change the division sign ($\div$) to a multiplication sign ($\times$).

$\frac{\overset{5}{\cancel{15}}}{\underset{4}{\cancel{16}}} \times \frac{\overset{1}{\cancel{4}}}{\underset{3}{\cancel{9}}} = \frac{5}{12}$

Step 3. Cancel: Divide 15 and 9 by 3. Cross out 15 and write 5. Cross out 9 and write 3. Divide 4 and 16 by 4. Cross out 4 and write 1. Cross out 16 and write 4.

Step 4. Multiply the numerators. $5 \times 1 = 5$

Step 5. Multiply the denominators. $4 \times 3 = 12$

Step 6. Try to reduce. $\frac{5}{12}$ is reduced.

Change every improper fraction answer to a whole or mixed number and reduce. Answers are on page 169.

1. $\frac{5}{6} \div 8\frac{1}{3} =$ $\frac{7}{12} \div 3\frac{1}{9} =$ $\frac{11}{15} \div 4\frac{2}{5} =$ $\frac{4}{9} \div 2\frac{2}{3} =$

2. $\frac{3}{8} \div 1\frac{4}{5} =$ $\frac{9}{10} \div 2\frac{2}{5} =$ $\frac{2}{3} \div 1\frac{5}{9} =$ $\frac{4}{7} \div 1\frac{2}{3} =$

3. $\frac{5}{6} \div 6\frac{1}{2} =$ $\frac{9}{14} \div 2\frac{4}{7} =$ $\frac{15}{16} \div 2\frac{1}{2} =$ $\frac{4}{9} \div 3\frac{1}{3} =$

**In these problems change each whole number to an improper fraction.
Put the whole number in the numerator. Put 1 in the denominator.**

4. $\quad 4 \div 1\frac{3}{5} =$ $\qquad 3 \div 4\frac{1}{2} =$ $\qquad 9 \div 3\frac{3}{4} =$ $\qquad 6 \div 1\frac{2}{7} =$

5. $\quad 14 \div 3\frac{1}{2} =$ $\qquad 5 \div 1\frac{3}{7} =$ $\qquad 6 \div 2\frac{2}{3} =$ $\qquad 7 \div 4\frac{1}{5} =$

6. $\quad 8 \div 1\frac{1}{3} =$ $\qquad 2 \div 1\frac{3}{4} =$ $\qquad 3 \div 1\frac{5}{7} =$ $\qquad 20 \div 3\frac{1}{5} =$

**In these problems change both mixed numbers to improper fractions.
Then invert the fraction on the right.**

7. $\quad 3\frac{3}{4} \div 2\frac{1}{2} =$ $\qquad 1\frac{7}{8} \div 2\frac{1}{4} =$ $\qquad 5\frac{1}{2} \div 2\frac{1}{16} =$ $\qquad 2\frac{5}{8} \div 1\frac{3}{4} =$

8. $\quad 6\frac{1}{2} \div 9\frac{3}{4} =$ $\qquad 4\frac{4}{5} \div 1\frac{1}{15} =$ $\qquad 8\frac{1}{3} \div 3\frac{1}{3} =$ $\qquad 3\frac{5}{9} \div 4\frac{4}{9} =$

9. $\quad 2\frac{2}{9} \div 5\frac{1}{3} =$ $\qquad 2\frac{2}{9} \div 1\frac{2}{3} =$ $\qquad 6\frac{1}{4} \div 2\frac{11}{12} =$ $\qquad 7\frac{1}{2} \div 3\frac{1}{8} =$

Dividing Fractions Applications

These problems give you a chance to apply your division of fractions skills. There is no key word like **of** to tell you to divide. But some types of problems usually mean to divide.

For example, you may be given the cost of several things and be asked to find the cost of one thing.

EXAMPLE: $1\frac{3}{4}$ pounds of apples cost $1.19. Find the cost of one pound of apples.

To find the cost of one pound, divide the cost of $1\frac{3}{4}$ pounds, $1.19, by $1\frac{3}{4}$. Remember that **the thing being divided must come first.** In this problem the money is being divided up into smaller amounts. The money comes first.

$$\$1.19 \div 1\frac{3}{4} = \frac{1.19}{1} \div \frac{7}{4} = \frac{\overset{17}{\cancel{1.19}}}{1} \times \frac{4}{\cancel{7}} = \$.68$$

In money problems be sure to keep the point that separates dollars and cents.

The words **cutting**, **sharing**, and **splitting** usually mean to divide.

Give each answer the correct label such as yards or cents. **Answers are on page 169.**

1. Harriet paid $4.05 for $2\frac{1}{4}$ pounds of ground beef. Find the cost of one pound of beef.

2. Carl wants to split a board $48\frac{3}{4}$ inches long into 6 equal pieces. How long will each piece be?

3. Jane is filling cans with $\frac{1}{2}$ pound of string beans. How many cans can she fill with 11 pounds of string beans?

4. Angelo wants to cut a piece of copper tubing $67\frac{1}{2}$ inches long into $7\frac{1}{2}$-inch pieces. How many pieces can he cut from the tube?

5. Three friends went fishing. They agreed to share the fish equally. Altogether they caught $15\frac{3}{4}$ pounds of fish. How many pounds did each person get?

6. Selma paid $16.50 for $3\frac{2}{3}$ yards of material. Find the price of one yard of the material.

7. In March the Rigbys made 9 long distance calls. The total time for the calls was $56\frac{1}{4}$ minutes. What was the average time for each call?

8. Mr. Brown is selling 16 acres of his farm to Mr. Smith. Mr. Smith plans to build houses on the land. Each house will be on a $\frac{1}{3}$-acre lot. How many lots can Mr. Smith get from Mr. Brown's land?

9. George needs $6\frac{1}{2}$ feet of lumber to build a bookcase. How many bookcases can he build from 39 feet of lumber?

10. June baked 9 pounds of cornbread. Each baking pan holds $\frac{3}{4}$ pound. How many pans did she use to bake all the cornbread?

11. Jim is building a brick walk from his driveway to his front door. Each brick is $3\frac{1}{2}$ inches wide. He wants the walk to be 49 inches wide. How many rows of bricks does he need?

Fraction Review

These problems will help you find out if you need to review the fraction section of this book. When you finish, look at the chart to see which pages you should review.

1. For each picture write a fraction that shows what part of the picture is shaded.

 ___   ___   ___   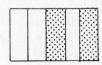 ___

2. There are 1000 meters in a kilometer. 13 meters are what fraction of a kilometer?

3. There are 24 hours in a day. 7 hours are what fraction of a day?

4. Circle the proper fractions in this list:

    $\frac{9}{9}$ $3\frac{1}{2}$ $\frac{4}{11}$ $\frac{9}{2}$ $\frac{3}{8}$ $\frac{6}{5}$

5. Circle the improper fractions in this list:

    $\frac{9}{2}$ $\frac{4}{7}$ $8\frac{1}{2}$ $\frac{2}{11}$ $\frac{6}{6}$ $\frac{13}{3}$

6. Circle the mixed numbers in this list:

    $\frac{3}{3}$ $\frac{2}{15}$ $\frac{15}{2}$ $\frac{1}{6}$ $8\frac{4}{7}$ $\frac{2}{9}$

7. Reduce each fraction to lowest terms.

    $\frac{7}{84} =$ $\frac{25}{45} =$ $\frac{32}{96} =$ $\frac{40}{500} =$

8. Raise each fraction to higher terms by finding the missing number.

    $\frac{9}{11} = \frac{}{33}$ $\frac{4}{7} = \frac{}{56}$ $\frac{8}{9} = \frac{}{54}$ $\frac{5}{16} = \frac{}{32}$

9. Change each fraction to a whole or mixed number. Reduce each fraction that is left.

    $\frac{40}{6} =$ $\frac{8}{8} =$ $\frac{36}{15} =$ $\frac{30}{9} =$

10. Change each mixed number to an improper fraction.

$6\frac{2}{7} =$ $\qquad$ $5\frac{3}{4} =$ $\qquad$ $4\frac{5}{6} =$ $\qquad$ $9\frac{1}{3} =$

11. Circle the bigger fraction in each pair.

$\frac{5}{8}$ or $\frac{7}{10}$ $\qquad$ $\frac{8}{25}$ or $\frac{1}{3}$ $\qquad$ $\frac{1}{6}$ or $\frac{3}{20}$ $\qquad$ $\frac{5}{9}$ or $\frac{2}{3}$

12. On Friday John had to drive 480 miles. He stopped to eat when he had driven 180 miles. What fraction of the trip had John finished when he stopped to eat?

13. Fran borrowed $3,000 to buy a new car. So far she has paid back $2,400. What fraction of the loan has she paid back?

14.
$$\frac{3}{7}$$
$$+\frac{2}{7}$$

15.
$$\frac{7}{10}$$
$$+\frac{1}{10}$$

16.
$$4\frac{5}{12}$$
$$+6\frac{11}{12}$$

17.
$$\frac{8}{15}$$
$$+\frac{4}{5}$$

18.
$$\frac{1}{2}$$
$$+\frac{5}{9}$$

19.
$$7\frac{2}{9}$$
$$8\frac{5}{6}$$
$$+3\frac{1}{4}$$

20. $6\frac{4}{5} + 9\frac{1}{2} + 2\frac{3}{4} =$

21. $3\frac{2}{3} + 4\frac{5}{8} + 5\frac{1}{4} =$

22. Cecilia bought $2\frac{1}{2}$ pounds of ground beef, $3\frac{5}{16}$ pounds of chicken, and $2\frac{7}{8}$ pounds of pork. Find the total weight of the things she bought.

23. The distance from Jack's house to his son's school is $4\frac{1}{2}$ miles. The distance from the school to Jack's factory is $5\frac{3}{10}$ miles. Find the distance from Jack's house to the factory by way of the school.

24.
$$\frac{11}{16}$$
$$-\frac{5}{16}$$

25.
$$\frac{13}{20}$$
$$-\frac{9}{20}$$

26.
$$\frac{3}{5}$$
$$-\frac{1}{6}$$

27.
$$9$$
$$-4\frac{5}{12}$$

28. $6\frac{5}{8}$ **29.** $8\frac{2}{5}$ **30.** $9\frac{1}{4} - 4\frac{7}{12} =$ **31.** $7\frac{1}{3} - 3\frac{5}{6} =$

$\quad\; -4\frac{7}{8}$ $\quad\; -2\frac{2}{3}$

32. The city of Weston hopes to raise $\$1\frac{3}{4}$ million to build a new school. So far it has $\$\frac{9}{10}$ million. How much more does it need?

33. Sam sawed a piece of lumber $15\frac{5}{8}$ inches long from a board 36 inches long. How long was the piece left over?

34. $\frac{3}{8} \times \frac{5}{7} =$ **35.** $\frac{2}{9} \times \frac{15}{16} =$ **36.** $\frac{5}{12} \times \frac{4}{5} =$ **37.** $8 \times \frac{7}{10} =$

38. $12 \times 1\frac{5}{6} =$ **39.** $\frac{4}{5} \times 1\frac{1}{8} =$ **40.** $1\frac{5}{7} \times 4\frac{2}{3} =$ **41.** $1\frac{7}{8} \times 3\frac{1}{5} =$

42. The Jordan family makes $12,900 a year. They spend $\frac{1}{3}$ of their income for food and clothing. How much do they spend for food and clothing in a year?

43. There are 624 employees of the city of Midvale. $\frac{5}{12}$ of them belong to a union. How many of the employees belong to a union?

44. $\frac{3}{10} \div \frac{2}{5} =$ **45.** $9 \div \frac{3}{4} =$ **46.** $3\frac{3}{4} \div \frac{5}{6} =$ **47.** $\frac{7}{8} \div 3 =$

48. $\dfrac{3}{5} \div 1\dfrac{5}{7} =$ **49.** $5 \div 3\dfrac{1}{3} =$ **50.** $1\dfrac{1}{3} \div 2\dfrac{2}{3} =$ **51.** $1\dfrac{7}{8} \div 6\dfrac{3}{4} =$

52. Paul paid $15.40 for $4\dfrac{2}{3}$ feet of lumber. How much did one foot of lumber cost?

53. Ann has 60 pounds of apples. She wants to put them in bags that hold $2\dfrac{1}{2}$ pounds. How many bags can she fill?

Check your answers on page 169. Then turn to the review pages for the problems you missed. Correct your answers before going on to the next page.

If you missed any of problems	review pages
1 to 3	59 to 60
4 to 6	61
7 to 10	62 to 66
11 to 13	67 to 69
14 to 16	70 to 71
17 to 21	72 to 75
22 to 23	76 to 77
24 to 26	78 to 79
27 to 31	80 to 82
32 to 33	83 to 84
34 to 36	85 to 87
37 to 41	88 to 90
42 to 43	91 to 92
44 to 47	93 to 96
48 to 51	97 to 98
52 to 53	99 to 100

Step One to Decimal Skill

These problems will help you find out if you need work in the decimal section of this book. Do all the problems you can. When you are finished, look at the chart to see which page you should go to next.

1. Write these decimals or mixed decimals in words.

    a. .5 _____   6.3 _____

    b. .07 _____   9.12 _____

    c. .035 _____   12.008 _____

    d. .0029 _____   3.000016 _____

2. Write each number as a decimal or a mixed decimal.

    a. nine tenths _____

    b. twenty-two thousandths _____

    c. sixteen ten-thousandths _____

    d. eighty-six hundred-thousandths _____

    e. five and two hundredths _____

    f. sixty and ninety-three thousandths _____

    g. four hundred seven and thirty-three millionths _____

3. Circle the bigger decimal in each pair:

    a. .092 or .9   .04 or .052   .304 or .04

    b. .3 or .298   .52 or .259   .07 or .081

4. Circle the biggest decimal in each group:

    a. .042, .4, or .04   .7, .707, or .06   .202, .02, or .22

    b. .82, .083, or .8   .05, .045, or .04   .49, .5, or .502

5. Change each decimal or mixed decimal to a fraction or mixed number. Reduce each fraction.

    a. .05 =   .9 =   3.8 =   6.375 =

    b. .048 =   6.065 =   .004 =   10.2 =

6. Change each fraction to a decimal.

    a. $\frac{4}{5} =$ $\frac{7}{12} =$ $\frac{11}{20} =$ $\frac{2}{7} =$

    b. $\frac{9}{10} =$ $\frac{12}{25} =$ $\frac{1}{9} =$ $\frac{9}{50} =$

7. .2 + .9 + .6 =   8. .58 + .9 + .737 =

9. .0046 + .53 + .094 =   10. .13 + 23.8 + 14 =

11. .68 + 10.24 + 5 =   12. .058 + 1.26 + 13 =

13. In 1970 there were 98.9 million males and 104.3 million females in the U.S. What was the total U.S. population in 1970?

14. Joe spent the weekend painting his house. Friday he worked 2.5 hours. Saturday he worked 6.25 hours. Sunday he worked 3.75 hours. How many hours did he spend painting his house?

15. .6 − .48 =   16. 3.4 − .75 =   17. 8.1 − .238 =

18. 12 − .608 =   19. 7.2 − 1.074 =   20. 16.48 − 3.3 =

21. Manny's empty suitcase weighs 1.2 kilograms. When he packed the suitcase, it weighed 10.15 kilograms. What was the weight of the things inside the suitcase?

22. In 1970 North America produced about 4.2 billion barrels of oil. In 1980 it produced 3.9 billion barrels. How much less oil did it produce in 1980?

23. $6 \times .35 =$

24. $4.3 \times 2.8 =$

25. $.26 \times 3.9 =$

26. $12.9 \times 5 =$

27. $.0038 \times 62 =$

28. $.417 \times 2.3 =$

29. Hannah bought 2.8 pounds of cheese. Each pound cost $2.40. How much did she pay altogether?

30. Bill makes $5.40 an hour. He works 37.5 hours a week. How much does Bill make each week?

31. $46.4 \div 16 =$

32. $.222 \div .6 =$

33. $.406 \div .07 =$

34. $121.9 \div 5.3 =$

35. $9 \div .18 =$

36. $148 \div 3.7 =$

37. Harry paid $13.05 for 4.5 feet of lumber. Find the price of one foot.

38. There are 1.6 kilometers in a mile. The distance from Sue's house to her job is 8 kilometers. Find the distance in miles.

Check your answers on page 170. Then complete the chart below.

Problem numbers	Number of problems in this section	Number of problems you got right in this section	
1 to 6	6	_____	If you had fewer than 4 problems right, go to page 109.
7 to 14	8	_____	If you had fewer than 6 problems right, go to page 114.
15 to 22	8	_____	If you had fewer than 6 problems right, go to page 117.
23 to 30	8	_____	If you had fewer than 6 problems right, go to page 119.
31 to 38	8	_____	If you had fewer than 6 problems right, go to page 122.

If you missed no more than 8 problems, correct then and go to Step One to Percent Skill on page 132.

Reading and Writing Decimals

A decimal is a kind of fraction. You have used decimals since you first handled money. $.60 is a decimal. It is 60 of the 100 equal parts in a dollar. When we write $.60 like a common fraction, it is $\frac{60}{100}$. Decimals are different from common fractions in two ways. One difference is that the denominators of decimals are not written. The other difference is that only some numbers — 10, 100, 1000, etc. — can be decimal denominators.

Decimals get their names from the number of **places** at the right of the decimal point. A place is the position of a digit. The decimal point itself does not take up a decimal place. The list below gives the names of the first six decimal places. Memorize this list before you go on. The list also gives examples of each of the decimals.

Number of places	Decimal		Example
one	tenths	.7	seven tenths
two	hundredths	.06	six hundredths
three	thousandths	.005	five thousandths
four	ten-thousandths	.0003	three ten-thousandths
five	hundred-thousandths	.00001	one hundred-thousandth
six	millionths	.000002	two millionths

Mixed decimals are numbers with digits on both sides of the decimal point. $9.48 is a mixed decimal. It means 9 whole dollars and $\frac{49}{100}$ of a dollar. Mixed decimals have whole numbers at the left of the decimal point.

Remember that a decimal gets its name from the number of places at the **right** of the decimal point. To read a decimal, count the places at the right of the point.

EXAMPLE: Read the decimal .042 .

Step 1. Count the decimal places: three. Three decimal places are thousandths.

Step 2. Read .042 as **forty-two thousandths.**

With mixed decimals, separate the whole number and the decimal with the word **and**.

EXAMPLE: Read 13.09 .

> *Step 1.* Count the decimal places: two. Two decimal places are hundredths. (13 is a whole number.)
>
> *Step 2.* Read 13.09 as **thirteen and nine hundredths.**

Write these decimals or mixed decimals in words. Answers are on page 171.

1. .3 _____   4.2 _____

2. .06 _____   8.07 _____

3. .015 _____   10.003 _____

4. .0016 _____   60.0007 _____

5. .00304 _____   3.000009 _____

When you write decimals, decide how many places you need. Use zeros in places that are not filled.

EXAMPLE: Write fourteen ten-thousandths as a decimal.

> *Step 1.* Decide how many places you need. Ten-thousandths need four places.

.0014 *Step 2.* The number 14 needs only two places. Use zeros in the first two places.

EXAMPLE: Write eleven and nine hundredths as a mixed decimal.

> *Step 1.* Write the whole number 11.
>
> *Step 2.* Decide how many decimal places you need. Hundredths need two places.
>
> *Step 3.* The digit 9 needs only one place. Use a zero in the first decimal place.

11.09 *Step 4.* Separate 11 and 09 with a decimal point.

Write each number as a decimal or mixed decimal. Answers are on page 171.

6. three tenths _____   five and four hundredths _____

7. thirteen thousandths _____   thirty and seven tenths _____

8. two hundredths _____   ninety and six thousandths _____

9. twelve ten-thousandths _____   sixteen millionths _____

10. eight and three hundred four millionths _____

Comparing Decimals

When you compare fractions, you first change the fractions to new fractions with a common denominator. When you compare decimals, first change the decimals to new decimals with the same number of places. Decimals with the same number of places have a common denominator. You can put zeros to the right of a decimal without changing its value. .5 and .50 have the same value. The 5 is in the tenths place in each decimal.

EXAMPLE: Which decimal is bigger, .08 or .6?

Step 1.   Put a zero at the right of .6 to change it to .60. Both decimals are hundredths now.

Step 2.   Decide which is bigger, .08 or .60 . Sixty hundredths is bigger than eight hundredths. .6 is the bigger decimal.

EXAMPLE: Which decimal is biggest, .34, .3, or .304?

Step 1.   Put a zero at the right of .34 to change it to .340.

Step 2.   Put two zeros at the right of .3 to change it to .300. All three decimals are thousandths now.

Step 3.   Decide which is biggest, .340, .300, or .304. Three hundred forty thousandths is the biggest. .34 is the biggest decimal.

Circle the bigger decimal in each pair. Answers are on page 171.

1.   .9 or .95   .27 or .3   .07 or .052

2.   .297 or .4   .004 or .04   .05 or .061

3.   .64 or .626   .33 or .323   .564 or .55

4.   .2 or .0347   .72 or .279   .08 or .1013

Circle the biggest decimal in each group.

5.   .7, .07, or .67   .407, .43, or .4   .0012, .201, or .12

6.   .29, .3, or .302   .5, .055, or .505   .707, .77, or .07

7.   .08, .028, or .82   .79, .097, or .709   .033, .03, or .3303

8.   .63, .6306, or .6   .09, .809, or .9   .2, .02, or .002

Changing Decimals to Fractions

A decimal is a kind of fraction. To change a decimal to a common fraction, write the digits in the decimal as the numerator. Write the name of the decimal (tenths, hundredths, thousandths, etc.) as the denominator. Then reduce the fraction.

EXAMPLE: Change .04 to a fraction.

$\underline{04}$ *Step 1.*   Write 04 as the numerator.

$\dfrac{04}{100}$ *Step 2.*   .04 has two decimal places. Two decimal places are hundredths. Write 100 as the denominator.

$\dfrac{04}{100} = \dfrac{1}{25}$ *Step 3.*   Reduce $\frac{04}{100}$ by 4. You do not need to write the zero in the numerator.

EXAMPLE: Change 5.8 to a mixed number.

$5\dfrac{8}{}$ *Step 1.*   Write 5 as a whole number. Write 8 has the numerator.

$5\dfrac{8}{10}$ *Step 2.*   5.8 has one decimal place. One decimal place is tenths. Write 10 as the denominator.

$5\dfrac{8}{10} = 5\dfrac{4}{5}$ *Step 3.*   Reduce $5\frac{8}{10}$ by 2.

Change each decimal or mixed decimal to a fraction or mixed number. Reduce each fraction. Answers are on page 171.

1. .08 =	.4 =	6.24 =	9.375 =
2. .625 =	.001 =	3.0025 =	7.42 =
3. .0055 =	.75 =	8.35 =	4.88 =
4. .48 =	.725 =	3.36 =	11.005 =
5. .035 =	.0075 =	6.064 =	2.00004 =

Changing Fractions to Decimals

A fraction means to divide. The line between the numerator and the denominator means **divided by**. For example, $\frac{3}{4}$ means 3 divided by 4. To change a fraction to a decimal, divide the numerator by the denominator. Put a decimal point and zeros to the right of the numerator.

EXAMPLE: Change $\frac{3}{4}$ to a decimal.

$$\begin{array}{r} .75 \\ 4\overline{)3.00} \\ \underline{2\ 8} \\ 20 \\ \underline{20} \end{array}$$

Step 1. Divide 3 by 4.

Step 2. Put a decimal point and two zeros to the right of 3.

Step 3. Divide and bring the decimal point up into the answer above its position in the problem.

Sometimes you can put just one zero to the right of the decimal point and the division will come out even. Sometimes you can keep putting zeros in the problem and the division will never come out even.

EXAMPLE: Change $\frac{1}{6}$ to a decimal.

$$\begin{array}{r} .16\frac{4}{6} \\ 6\overline{)1.00} \\ \underline{6} \\ 40 \\ \underline{36} \\ 4 \end{array}$$

Step 1. Divide 1 by 6.

Step 2. Put a decimal point and two zeros to the right of 1.

Step 3. Divide and bring the decimal point up into the answer above its position in the problem. Notice that if we put more zeros in the problem, the division does not come out even. You can stop with two zeros.

$.16\frac{4}{6} = .16\frac{2}{3}$ *Step 4.* Make a fraction with the remainder 4 over the divisor 6.

Step 5. Reduce the fraction by 2.

Change each fraction to a decimal. Answers are on page 171.

1. $\frac{1}{4} =$ $\frac{3}{5} =$ $\frac{7}{10} =$ $\frac{2}{3} =$ $\frac{5}{6} =$

2. $\frac{9}{20} =$ $\frac{3}{20} =$ $\frac{3}{8} =$ $\frac{16}{25} =$ $\frac{2}{9} =$

3. $\frac{3}{10} =$ $\frac{5}{12} =$ $\frac{4}{7} =$ $\frac{1}{2} =$ $\frac{4}{25} =$

4. $\frac{1}{3} =$ $\frac{2}{5} =$ $\frac{5}{8} =$ $\frac{9}{16} =$ $\frac{1}{12} =$

Addition of Decimals

Adding decimals is one of the easiest operations in math. To add decimals, line up the decimals with the decimal points under each other.

EXAMPLE: Add .63 + .4 + .279 =

```
  .63
  .4
+ .279
-------
1.309
```

Step 1. Line up the decimals with the points under each other.

Step 2. Add each column. In the thousandths column the only digit is 9. In the hundredths column the digits are 3 and 7. In the tenths column the digits are 6, 4, and 2.

In this example the total of the tenths column is 13. Only one digit can fit under each column. Write the 3 in the tenths place and carry the 1 over to the units column.

Add each problem. Answers are on page 171.

1. .28 + .3 + .709 =   .34 + .959 + .6 =

2. .3 + .8 + .6 =   .27 + .94 + .08 =

3. .68 + .7 + .697 =   .3 + .4177 + .274 =

4. .08 + .703 + .4 =   .288 + .8 + .0806 =

5. .963 + .0089 + .05 =   .5437 + .06 + .192 =

6. .007 + .36 + .4 =   .046 + .228 + .307 =

When you add whole numbers with decimals or mixed decimals, remember to put a point at the right of the whole number. Then line up the numbers with the decimal points under each other.

EXAMPLE: Add $4.73 + 29 + .586 =$

$$
\begin{array}{r}
4.73 \\
29. \\
+ \quad .586 \\
\hline
34.316
\end{array}
$$

Step 1.   Put a decimal point at the right of 29.

Step 2.   Line up the numbers with the points under each other.

Step 3.   Add each column.

7. $2.1 + 66 + 3.97 =$ $.506 + 5.6 + 4 =$

8. $70 + 6.256 + .49 =$ $5 + .92 + .747 =$

9. $.428 + 5 + .93 =$ $4.77 + .53 + 30 =$

10. $2.587 + 3.94 + 6 =$ $26.073 + 18 + .37 =$

11. $42 + 28.5 + 7.57 =$ $.075 + 2.8 + 6 =$

12. $3.0045 + 4 + .39 =$ $8 + .0682 + 2.7 =$

13. $249.1 + 3.772 + 8 =$ $.609 + 4.44 + 23 =$

Adding Decimals Applications

These problems give you a chance to apply your addition of decimals skills. For each problem give your answer the correct label such as pounds or miles. **Answers are on page 172.**

1. The average April temperature in Chicago is 47.8°. The average April temperature in St. Louis is 8.3° higher than in Chicago. Find the average April temperature for St. Louis.

2. In 1975 the town of Elmford spent $2.2 million for education. In 1980 it spent $.85 million more. How much did it spend on education in 1980?

3. On Monday Ann drove 3.7 miles to take her children to school. She also drove 1.9 miles to a gas station, 4.6 miles to a shopping center, and 5.8 miles back home. How many miles did she drive in total?

4. In 1970 there were about 3.7 billion people in the world. In 1980 there were about .8 billion more people. What was the world population in 1980?

5. The reading on the mileage gauge of Pete's car was 36,405.2 miles on Friday morning. By Sunday night Pete had driven 768.9 more miles. What was the reading Sunday night?

6. Rachel's normal temperature is 98.6°. When she was ill, her temperature went up 4.9°. What was her temperature when she was ill?

7. Jack works part-time at a garage. Monday he worked 4.5 hours. Wednesday he worked 3.25 hours. Friday he worked 5 hours. How many hours did he work altogether that week?

Subtraction of Decimals

To subtract decimals line up the decimals with the points under each other just like addition. Remember to put a point at the right of a whole number. Put zeros at the right until each decimal has the same number of places. You will need the zeros for borrowing.

EXAMPLE: Subtract 5 − .38 =

$$
\begin{array}{r}
5. \\
- \ .38 \\
\hline
\end{array}
$$

Step 1. Put a decimal point at the right of 5.

Step 2. Line up the numbers with the points under each other.

$$
\begin{array}{r}
4 \ \overset{9}{\cancel{10}} \ \cancel{10} \\
\cancel{5}.\cancel{0}\cancel{0} \\
- \ .38 \\
\hline
4.62
\end{array}
$$

Step 3. Put two zeros at the right of 5 to give each decimal the same number of places.

Step 4. Borrow and subtract.

Subtract each problem. Answers are on page 172.

1. 6 − .359 =	12 − .35 =	.7 − .482 =
2. .38 − .098 =	4 − .059 =	.02 − .004 =
3. 3.8 − 2.947 =	5.8 − .399 =	.63 − .406 =
4. 20 − 3.89 =	6 − 2.075 =	8.76 − 3 =
5. .07 − .052 =	30 − .8 =	.3 − .049 =
6. .036 − .009 =	2 − .4306 =	15 − .8 =
7. 9.6 − 1.81 =	.5 − .279 =	4 − .404 =

Subtracting Decimals Applications

These problems give you a chance to apply your subtraction of decimals skills. Remember to put the bigger number on top. Give your answers the correct labels such as meters or pounds. **Answers are on page 172.**

1. The area of the United States is about 3.3 million square miles. The area of Canada is about 3.8 million square miles. How much bigger is Canada in area?

2. There are about 222.1 million people living in the U.S. About 23.4 million people live in Canada. How many more people live in the U.S. than in Canada?

3. When George bought his used car the mileage gauge read 15,023.4 miles. In two months the gauge read 19,376.8 miles. How many miles did George drive the first two months?

4. Ty Cobb's batting average for his career was .367. Rogers Hornsby's average was .358. How much better was Ty Cobb's average?

5. Judith is 1.6 meters tall. Her daughter Emma is 1.35 meters tall. How much taller is Judith than her daughter?

6. In 1970 there were 6.7 million people living in the Chicago area. In 1980 there were 7.5 million in the area. By how much did the population grow from 1970 to 1980?

7. Colin bought a piece of lumber 2 meters long. From it he cut a piece 1.85 meters long. How long was the piece left over?

8. Mary bought 12.4 pounds of apples. She gave 1.9 pounds of apples to her sister. How many pounds of apples were left?

Multiplication of Decimals

Multiplying decimals is easier than adding or subtracting decimals. You do not have to line up the decimals. Just count the number of decimal places in each number you multiply. Put the total number of places from the two numbers in the answer.

EXAMPLE: Multiply 2.36 × .4 =

Step 1. Set the numbers up for easy multiplication. .4 has only one digit. Put it below.

Step 2. Multiply the numbers.

$$\begin{array}{r} 2.36 \\ \times \quad .4 \\ \hline .9\,44 \end{array}$$

Step 3. Count the number of decimal places in each number. 2.36 has two decimal places. .4 has one decimal place.

Step 4. Put the total number of decimal places (2 + 1 = 3) in the answer.

Remember that a whole number has no decimal places.

EXAMPLE: Multiply 5 × 6.28 =

Step 1. Set the numbers up for easy multiplication.

Step 2. Multiply the numbers.

$$\begin{array}{r} 6.28 \\ \times \quad 5 \\ \hline 25.4\cancel{0} \end{array}$$

25.4

Step 3. Count the number of decimal places in each number. 6.28 has two decimal places. 5 has no decimal places.

Step 4. Put the total number of decimal places (2 + 0 = 2) in the answer.

Step 5. Take off the last zero from the answer. The zero is not necessary. It does not change the value of the other digits.

Sometimes you will need to put extra zeros in your answer.

EXAMPLE: Multiply .3 × .07 =

Step 1. Set the numbers up for easy multiplication.

Step 2. Multiply the numbers.

$$\begin{array}{r} .07 \\ \times \quad .3 \\ \hline .021 \end{array}$$

Step 3. Count the number of decimal places in each number. .3 has one. .07 has two.

Step 4. Put the total number of decimal places (1 + 2 = 3) in the answer. Put a zero at the left of 21 to make three places.

Multiply each problem. Answers are on page 172.

1. $7 \times 2.6 =$ $.82 \times 9 =$ $8 \times .03 =$

2. $.09 \times .8 =$ $.6 \times 9.1 =$ $.44 \times .5 =$

3. $.6 \times 8.23 =$ $.974 \times .7 =$ $.4 \times 83.7 =$

4. $.23 \times 82 =$ $1.9 \times 8.6 =$ $.96 \times 3.3 =$

5. $.047 \times .4 =$ $.006 \times .07 =$ $.36 \times .004 =$

6. $348 \times .05 =$ $4.7 \times 296 =$ $352 \times .16 =$

7. $2.5 \times 4.16 =$ $.81 \times 60.7 =$ $3.8 \times .529 =$

8. $1.29 \times 30 =$ $60 \times .408 =$ $87.3 \times 70 =$

9. $.048 \times .66 =$ $817 \times .05 =$ $.0635 \times .4 =$

Multiplying Decimals Applications

These problems give you a chance to apply your multiplication of decimals skills. Give every answer the correct label such as grams or dollars. **Answers are on page 172.**

1. Jeff weighs 180 pounds. One pounds equals 0.45 kilogram. What is Jeff's weight in kilograms?

2. Adrienne works overtime for $8.60 an hour. Last week she worked 7.5 hours overtime. How much did she make for overtime last week?

3. Tom's car gets an average of 14.5 miles of city driving with one gallon of gas. How many miles of driving can he get in the city with 12.4 gallons of gas?

4. Tom pays $1.35 for a gallon of gas. How much does he have to pay for 12.4 gallons?

5. Ruby bought chicken for $1.40 a pound. How much did she pay for 2.3 pounds of chicken?

6. Renee is 64 inches tall. One inch equals 2.54 centimeters. Find Renee's height in centimeters.

7. Fred walks at an average speed of 3.6 miles per hour. How far can he walk in 2.5 hours?

8. It costs 2.5¢ to run a black and white T.V. for an hour. How much does it cost to run a T.V. for 8 hours?

9. Mark bought 4.25 feet of lumber. The lumber cost $3.40 a foot. What was the total cost of the lumber?

Division of Decimals by Whole Numbers

You began to divide decimals when you changed fractions into decimals on page 113. There you divided a whole number by a bigger whole number. When you divide a decimal by a number, line up your problem carefully. Then bring the decimal point up into the answer above its position in the problem.

EXAMPLE: Divide 4.68 ÷ 6 =

```
   .78
6)4.68
  4 8
    56
    56
```

Step 1. Set the problem up for long division.

Step 2. Divide.

Step 3. Bring the decimal point up into the answer above its position in the problem.

Sometimes you will need to put zeros in your answer.

EXAMPLE: Divide .512 ÷ 8 =

```
   .064
8).512
   48
    32
    32
```

Step 1. Set the problem up for long division.

Step 2. Divide.

Step 3. Bring the decimal point up into the answer above its position in the problem. To show that 8 does not divide into .5, put a zero above the 5.

Divide each problem. Answers are on page 172.

1. 29.6 ÷ 8 =   3.12 ÷ 6 =   .342 ÷ 9 =

2. 13.16 ÷ 4 =   4.368 ÷ 7 =   2.01 ÷ 3 =

3. .324 ÷ 12 =   67.2 ÷ 24 =   .702 ÷ 18 =

4. 604.5 ÷ 15 =   1.633 ÷ 23 =   26.04 ÷ 42 =

Division of Decimals by Decimals

Dividing by decimals is the most complicated decimal operation. First change the problem to a new problem. In the new problem the number you are dividing by (the divisor) should be a whole number. You can change the divisor to a whole number by moving the decimal point to the right end. Then move the decimal point in the other number (the dividend) the same number of places.

These steps are easier to understand with whole numbers. Look at the problem $10 \div 2 = 5$. The answer is the same if we move the decimal point one place to the right in both 10 and 2: $100 \div 20 = 5$. The problems are different, but the answers are the same.

With decimal division you can also change the problem and get the right answer. The object is always to make the divisor a whole number.

EXAMPLE: Divide $3.68 \div .8 =$

.8)3.68   *Step 1.*   Set the problem up for long division.

.8)3.68   *Step 2.*   Move the decimal point in the divisor .8 one place to the right to make it a whole number.

4.6
.8)3.6 8   *Step 3.*   Also move the decimal point in the dividend 3.68 one place to the right.
  3 2
  —————   *Step 4.*   Divide and bring the decimal point up into the answer above its new position.
    4 8
    4 8

Sometimes you will have to put extra zeros with the dividend.

EXAMPLE: Divide $5.6 \div .07 =$

.07)5.6   *Step 1.*   Set the problem up for long division.

.07)5.6   *Step 2.*   Move the decimal point in the divisor .07 two places to the right to make it a whole number.

    80.
.07)5.60   *Step 3.*   Also move the decimal point in the dividend 5.6 two places to the right. Put an extra zero to the right of 5.6 to get two decimal places.
  5 6
  ————   *Step 4.*   Divide and bring the decimal point up into the
    00   answer above its new position.

Divide each problem. Answers are on page 172.

1. $2.38 \div .7 =$ $.504 \div .9 =$ $2.34 \div .3 =$

2. $.282 \div .6 =$ $7.36 \div .8 =$ $22.8 \div .4 =$

3. $.0185 \div .05 =$ $.567 \div .09 =$ $.0348 \div .06 =$

4. $.0378 \div .07 =$ $.072 \div .04 =$ $7.44 \div .08 =$

5. $2.82 \div .006 =$ $.078 \div .003 =$ $.531 \div .009 =$

6. $11.88 \div .18 =$ $3.234 \div .42 =$ $.1728 \div .27 =$

7. $.333 \div 3.7 =$ $148.5 \div 4.5 =$ $552 \div 9.2 =$

8. $8.502 \div .026 =$ $1.0388 \div .053 =$ $.23485 \div .061 =$

9. $4.5955 \div 5.05 =$ $37.312 \div 70.4 =$ $40.45 \div .809 =$

Division of Whole Numbers by Decimals

To divide a whole number by a decimal, remember to put a decimal point at the right of the whole number. Then move the points in both the divisor and the dividend. You will have to put zeros with the dividend.
Follow this example carefully.

EXAMPLE: Divide 18 ÷ .003 =

.003⟌18.

Step 1. Set the problem up for long division. Put a decimal point at the right of 18.

.003⟌18.

Step 2. Move the decimal point in the divisor .003 three places to the right to make it a whole number.

.003⟌18.000
       18

0 000

Step 3. Also move the decimal point in the dividend three places to the right. Put three zeros to the right of 18 to get three decimal places.

Step 4. Divide and bring the decimal point up into the answer above its new position.

Divide each problem. Answers are on page 173.

1. 36 ÷ 2.4 =   30 ÷ .75 =   39 ÷ .06 =

2. 54 ÷ 4.5 =   20 ÷ .08 =   18 ÷ .036 =

3. 66 ÷ .12 =   3 ÷ .075 =   32 ÷ .64 =

4. 9 ÷ 2.25 =   147 ÷ .035 =   168 ÷ 2.8 =

Dividing Decimals Applications

These problems give you a chance to apply your division of decimals skills. Remember that **the number being divided goes inside the** ⌐ **sign.**

EXAMPLE: Linda bought 3.5 pounds of chicken for $3.43.
Find the cost of one pound of chicken.

3.5)$3.43 *Step 1.* Divide the price of several pounds ($3.43) by the number of pounds (3.5). In this problem the money is being divided into parts. Put the money inside the ⌐ sign.

3.5)$3.43 *Step 2.* Move the decimal point in the divisor one place to the right to make it a whole number.

```
        $   .98
3.5)$3.4 30
     3 1 5
     ────
     2 80
     2 80
```
Step 3. Also move the decimal point in the dividend one place to the right.

Step 4. Divide and bring the decimal point up into the answer above its new position. Put an extra zero at the right of 34.3 to give the answer the two decimal places for money.

Give each answer the correct label such as $ or meters. Answers are on page 173.

1. Allen bought 5.5 feet of lumber for $14.85. What was the price of one foot of lumber?

2. Jake is a plumber. He wants to cut a piece of pipe 2.52 meters long into 4 equal pieces. How long will each piece be?

3. There are 2.2 pounds in one kilogram. Verva weighs 154 pounds. What is her weight in kilograms?

126

4. There are 3.8 liters in a gallon. How many gallons are there in 95 liters?

5. Petra sent a package weighing 17 pounds. She paid $7.31 to send the package. What was the shipping price for sending each pound?

6. John works 35 hours a week. In a week he makes $162.75 before taxes. How much does John make in one hour?

7. Debbie drove 270 miles with 12.5 gallons of gas. How far did she drive with one gallon of gas?

8. A box holds 16 cans of soup. The total weight of all the cans in the box is 5.536 kilograms. How much does one can of soup weigh?

9. The Smiths picked 38.4 pounds of peaches. They want to share the peaches equally with three other families and themselves. How many pounds will each of the 4 families get?

10. There are 1.6 kilometers in a mile. How many miles are there in 88 kilometers?

11. Bob is a carpenter. He wants to split a board 2.45 meters long into 5 equal pieces. How long will each piece be?

Decimal Review

These problems will help you find out if you need to review the decimal section of this book. When you finish, look at the chart to see which pages you should review.

1. Write these decimals or mixed decimals in words.

   a. .6 _____   8.1 _____

   b. .13 _____   2.05 _____

   c. .004 _____   9.108 _____

   d. .0006 _____   7.0052 _____

2. Write each number as a decimal or a mixed decimal.

   a. four tenths _____

   b. eighteen thousandths _____

   c. five ten-thousandths _____

   d. thirteen millionths _____

   e. one and eight hundredths _____

   f. eighty-five and fifty-six thousandths _____

   g. two hundred twenty and six ten-thousandths _____

3. Circle the bigger decimal in each pair:

   a. .5 or .48   .029 or .2   .39 or .9

   b. .52 or .504   .06 or .6   .25 or .052

4. Circle the biggest decimal in each group:

   a. .78, .708, or .87   .101, .11, or .1   .7, .679, or .79

   b. .004, .041, or .04   .28, .082, or .208   .43, .406, or .4

5. Change each decimal or mixed decimal to a fraction or mixed number. Reduce each fraction.

   a. .3 =   3.75 =   4.16 =   .0025 =

   b. .0015 =   2.06 =   .625 =   9.32 =

6. Change each fraction to a decimal.

    a. $\dfrac{11}{12} =$ $\dfrac{1}{5} =$ $\dfrac{1}{8} =$ $\dfrac{4}{25} =$

    b. $\dfrac{3}{50} =$ $\dfrac{8}{9} =$ $\dfrac{1}{16} =$ $\dfrac{1}{7} =$

7. .3 + .5 + .7 =   **8.** .29 + .8 + .626 =

9. .0052 + .84 + .072 =   **10.** .26 + 14.7 + 13 =

11. .88 + 12.07 + 3 =   **12.** 18 + .049 + 2.38 =

13. The steps in front of Richard's house were 4 feet wide. Richard built an extension to make them 3.25 feet wider. How wide are the steps now?

14. In 1970 there were 8.4 million people in the Los Angeles area. In 1980 there were 2.3 million more people. How many people lived in the Los Angeles area in 1980?

15. .8 − .29 =   **16.** 4.5 − .83 =   **17.** 6.2 − .127 =

18. 11 − .509 =   **19.** 8.3 − 2.052 =   **20.** 3.24 − .966 =

21. Jorge is 1.9 meters tall. His son Mateo is 1.05 meters tall. How much taller is Jorge?

22. The average man is expected to live 68.1 years. The average woman is expected to live 75.4 years. On the average how much longer do women live?

23. $.47 \times 9 =$

24. $3.4 \times 1.9 =$

25. $4.3 \times .38 =$

26. $6 \times 21.7 =$

27. $18 \times .0074 =$

28. $6.5 \times .329 =$

29. Jenny makes $4.90 an hour. How much does she make on a day when she works 8.5 hours?

30. Joel drives 12 miles to get to work. One mile equals 1.6 kilometers. Find the distance to Joel's work in kilometers.

31. $110.4 \div 23 =$

32. $.296 \div .8 =$

33. $.621 \div .09 =$

34. $.3108 \div 4.2 =$

35. $54 \div .27 =$

36. $138 \div 4.6 =$

37. Jane paid $4.48 for 2.8 pounds of ground beef. Find the price of one pound.

38. Joe wants to share 10.8 pounds of fish with himself and 2 friends. How many pounds will each person get?

Check your answers on page 173. Then turn to the review pages for the problems you missed. Correct your answers before going on to the next page.

If you missed problems	review pages
1 to 2	109 to 110
3 to 6	111 to 113
7 to 14	114 to 116
15 to 22	117 to 118
23 to 30	119 to 121
31 to 38	122 to 127

Step One to Percent Skill

These problems will help you find out if you need work in the percent section of this book. Do all the problems you can. When you are finished, look at the chart to see which page you should go to next.

1. Change each decimal to a percent.

$.6 =$ $\qquad$ $.06 =$ $\qquad$ $.248 =$ $\qquad$ $.03\frac{1}{3} =$

2. Change each percent to a decimal.

$50\% =$ $\qquad$ $7\% =$ $\qquad$ $5\frac{1}{4}\% =$ $\qquad$ $325\% =$

3. Change each fraction to a percent.

$\frac{3}{10} =$ $\qquad$ $\frac{1}{8} =$ $\qquad$ $\frac{3}{7} =$ $\qquad$ $\frac{4}{25} =$

4. Change each percent to a fraction.

$36\% =$ $\qquad$ $8.5\% =$ $\qquad$ $31\frac{1}{4}\% =$ $\qquad$ $62\frac{1}{2}\% =$

5. 15% of 140 =

6. 90% of 60 =

7. 225% of 48 =

8. 10.5% of 200 =

9. $12\frac{1}{2}\%$ of 136 =

10. $83\frac{1}{3}\%$ of 72 =

11. Harry took a test with 60 questions. He got 85% of the questions right. How many questions did he get right?

12. Kate bought a radio on sale. The radio used to cost $34. It was on sale for 15% off. How much did Kate pay for the radio?

13. Find the interest on $1,800 at 8% annual interest for one year.

14. Find the interest on $920 at 6% annual interest for one year and 3 months.

15. 72 is what % of 90?

16. 44 is what % of 132?

17. 105 is what % of 140?

18. 120 is what % of 144?

19. There are 45 bus drivers in Midvale. 9 of the bus drivers are women. What percent of the drivers are women?

20. Last year the Gomez family paid $175 a month for rent. This year they have to pay $196 a month. By what percent did their rent go up?

21. 90% of what number is 54?

22. 15% of what number is 39?

23. 25% of what number is 35?

24. $16\frac{2}{3}$% of what number is 25?

25. 20% of the students in Mr. Green's math class were absent because of a snowstorm on Thursday. 5 students were absent. How many students are usually in the class?

26. Sophie got a 12% raise. Her raise amounts to $22.20 a week. How much was she making each week before the raise?

Check your answers on page 173. Then complete the chart below.

Problem numbers	Number of problems in this section	Number of problems you got right in this section	
1 to 4	4	_____	If you had fewer than 3 problems right, go to page 135.
5 to 14	10	_____	If you had fewer than 8 problems right, go to page 140.
15 to 20	6	_____	If you had fewer than 4 problems right, go to page 147.
21 to 26	6	_____	If you had fewer than 4 problems right, go to page 151.

Percents and Decimals

You already learned that a decimal is a kind of fraction. A percent is also a kind of fraction. Percents are used in everyday business problems. Interest rates, sales tax, and discounts are all measured in percents. For example, 6% is the sales tax rate in some states. 6% means 6 out of 100 equal parts. For every 100 cents you spend on something, you pay 6 cents in tax.

Percents are almost like two-place decimals. A decimal with two places is called hundredths. The denominator for percents is 100. But, instead of two decimal places, percents use the % sign.

It is easy to change decimals to percents. Move the decimal point two places to the **right** and write a percent sign.

EXAMPLE: Change .35 to a percent.

$.35 = .35 = 35\%$ *Step 1.* Move the point two places to the right.

                     Step 2. Write a percent sign.

Notice that when the decimal point moves to the end, you do not have to write it.

EXAMPLE: Change .052 to a percent.

$.052 = .05\,2 = 5.2\%$ *Step 1.* Move the point two places to the right.

                     Step 2. Write a percent sign.

EXAMPLE: Change $.66\frac{2}{3}$ to a percent.

$.66\frac{2}{3} = .66\frac{2}{3} = 66\frac{2}{3}\%$ *Step 1.* Move the point two places to the right.

                     Step 2. Write a percent sign.

Notice that when the decimal point comes just before a fraction, you do not have to write it.

Sometimes you will have to put zeros after the decimal to get two places.

EXAMPLE: Change .7 to a percent.

                     Step 1. Put a zero at the right of .7

$.7 = .70 = 70\%$ *Step 2.* Move the point two places to the right.

                     Step 3. Write a percent sign.

Change each decimal to a percent. Answers are on page 174.

1. .65 =   .06 =   .045 = $.16\frac{2}{3} =$

2. .8 =   .25 = $.06\frac{1}{4} =$ .82 =

3. .004 =   .2 =   .03 =   .36 =

4. 2.8 =   .5 =   .625 =   4. =

To change a percent to a decimal, move the decimal point two places to the **left.** Then take off the percent sign.

EXAMPLE: Change 65% to a decimal.

65% = ‿65 % = .65   *Step 1.* Move the point two places to the left.

Step 2. Take off the percent sign.

Sometimes you will have to put zeros at the left to get two places.

EXAMPLE: Change 2.8% to a decimal.

Step 1. Put a zero to the left of 2.8%.

2.8% = ‿02 8% = .028   *Step 2.* Move the point two places to the left.

Step 3. Take off the percent sign.

EXAMPLE: Change 20% to a decimal.

20% = ‿20 % = .2   *Step 1.* Move the point two places to the left.

Step 2. Take off the percent sign.

Notice that we took off the zero at the right in .20. The zero does not change the value of .2.

EXAMPLE: Change $44\frac{4}{9}$% to a decimal.

$44\frac{4}{9}$ % = ‿$44\frac{4}{9}$ % = $.44\frac{4}{9}$ *Step 1.* The point is understood to be at the right of 44. Move the point two places to the left.

Step 2. Take off the percent sign.

Change each percent to a decimal. Answers are on page 174.

5. 55% =   8% =   12.5% =   2% =

6. 6.4% = $33\frac{1}{3}$% =   60% =   225% =

7. 90% =   0.4% =   20% =   500% =

8. $62\frac{1}{2}$% =   300% = $1\frac{2}{3}$% =   11.4% =

Percents and Fractions

Percents are different from common fractions in two ways. One difference is that 100 is the only number that can be a denominator for percents. The other difference is that the denominator is not written. Instead of writing 100, we write the % sign. There are two different ways to change a fraction to a percent. Look at the examples on this page carefully. Then choose the way you like better.

One method for changing a fraction to a percent is to multiply the fraction by 100.

EXAMPLE: Change $\frac{2}{5}$ to a percent.

$$\frac{2}{5} \times 100 = \frac{2}{\underset{1}{\cancel{5}}} \times \frac{\overset{20}{\cancel{100}}}{1} = \frac{40}{1} = 40\%$$

When you use this method, remember to add the percent sign to the answer.

The other method is to change the fraction to a decimal first. Then change the decimal to a percent.

EXAMPLE: Change $\frac{3}{4}$ to a percent.

$\frac{3}{4} = 4\overline{)3.00}^{.75}$ *Step 1.* Change $\frac{3}{4}$ to a decimal. (Look at page 113).

$.75 = .75 = 75\%$ *Step 2.* Change .75 to a percent.

Change each fraction to a percent. Answers are on page 174.

1. $\frac{7}{10} =$ $\frac{3}{5} =$ $\frac{1}{9} =$ $\frac{9}{50} =$

2. $\frac{8}{25} =$ $\frac{3}{8} =$ $\frac{1}{3} =$ $\frac{1}{2} =$

3. $\frac{1}{6} =$ $\frac{2}{7} =$ $\frac{19}{100} =$ $\frac{5}{8} =$

4. $\frac{5}{12} =$ $\frac{9}{20} =$ $\frac{1}{16} =$ $\frac{1}{4} =$

Remember that a percent is a kind of fraction. To change a percent to a fraction, write the digits in the percent as the numerator. Write 100 as the denominator. Then reduce the fraction.

EXAMPLE: Change 85% to a fraction.

$\frac{85}{100} = \frac{17}{20}$ *Step 1.* Write 85 as the numerator and 100 as the denominator.

             Step 2. Reduce the fraction by 5.

When a percent has a decimal in it, first change the percent to a decimal. Then change the decimal to a fraction and reduce.

EXAMPLE: Change 8.4% to a fraction.

8.4% = .08 4 = .084   *Step 1.* Change 8.4% to a decimal.

$\frac{84}{1000} = \frac{21}{250}$ *Step 2.* Change .084 to a fraction.

Step 3. Reduce the fraction by 4.

When a percent has a fraction in it, write the digits in the percent as the numerator. Write 100 as the denominator. Then **divide** the numerator by the denominator. This operation is hard. Study the following example carefully.

EXAMPLE: Change $58\frac{1}{3}$% to a fraction.

$\frac{58\frac{1}{3}}{100}$ *Step 1.* Write $58\frac{1}{3}$ as the numerator and 100 as the denominator.

$58\frac{1}{3} \div 100 =$ *Step 2.* Remember that the line separating the numerator from the denominator means **divided by**. Divide $58\frac{1}{3}$ by 100.

$\frac{175}{3} \div \frac{100}{1} =$

$\frac{\overset{7}{\cancel{175}}}{3} \times \frac{1}{\underset{4}{\cancel{100}}} = \frac{7}{12}$

Change each percent to a fraction and reduce. Answers are on page 174.

5. 35% =	2% =	24% =	30% =
6. 44% =	6% =	150% =	3% =
7. 16% =	450% =	15% =	96% =
8. 4.8% =	10.5% =	.04% =	2.75% =
9. $12\frac{1}{2}$% =	$83\frac{1}{3}$% =	$42\frac{6}{7}$% =	$8\frac{1}{3}$% =
10. $55\frac{5}{9}$% =	$87\frac{1}{2}$% =	$6\frac{1}{4}$% =	$16\frac{2}{3}$% =

Common Fractions, Decimals, and Percents

The chart on this page includes some of the fractions, decimals, and percents you will use most often in your work. You will learn that sometimes it is easier to change a percent to a decimal. Sometimes it is easier to change a percent to a fraction. Fill in the chart. Then check your answers on page 174. Make sure your answers are correct. Then study the list carefully. Memorize the fraction and decimal each percent is equal to.

PERCENT	DECIMAL	FRACTION	PERCENT	DECIMAL	FRACTION
50%	_____	_____	20%	_____	_____
			40%	_____	_____
25%	_____	_____	60%	_____	_____
75%	_____	_____	80%	_____	_____
$12\frac{1}{2}\%$	_____	_____	10%	_____	_____
$37\frac{1}{2}\%$	_____	_____	30%	_____	_____
$62\frac{1}{2}\%$	_____	_____	70%	_____	_____
$87\frac{1}{2}\%$	_____	_____	90%	_____	_____
$33\frac{1}{3}\%$	_____	_____	$16\frac{2}{3}\%$	_____	_____
$66\frac{2}{3}\%$	_____	_____	$83\frac{1}{3}\%$	_____	_____

Finding a Percent of a Number

When you studied fractions you learned that a fraction immediately followed by the word **of** means to multiply. A percent immediately followed by the word **of** also means to multiply. To multiply by a percent, first change the percent to a fraction or a decimal. Then multiply by the fraction or decimal.

Sometimes fractions are easier to use. Sometimes decimals are easier to use. Study these examples carefully.

EXAMPLE: Find 40% of 65.

Using a fraction:

$40\% = \dfrac{40}{100} = \dfrac{2}{5}$　　　　*Step 1.* Change 40% to a fraction.

$\dfrac{2}{5} \times 65 =$　　　　*Step 2.* Multiply 65 by $\frac{2}{5}$.

$\dfrac{2}{\underset{1}{5}} \times \dfrac{\overset{13}{\cancel{65}}}{1} = \dfrac{26}{1} = 26$

Using a decimal:

$40\% = 40\% = .4$　　*Step 1.* Change 40% to a decimal.

$\begin{array}{r} 65 \\ \times\quad .4 \\ \hline 26.0 = 26 \end{array}$　　*Step 2.* Multiply 65 by .4.

In the example above both the fraction and the decimal are easy to use. When the percent itself has a decimal, it is easier to change the percent to a decimal.

EXAMPLE: Find 8.5% of 20.

$8.5\% = 08.5\% = .085$　　*Step 1.* Change 8.5% to a decimal.

$\begin{array}{r} 20 \\ \times\ .085 \\ \hline 100 \\ 1\ 60\quad \\ \hline 1.700 = 1.7 \end{array}$　　*Step 2.* Multiply 20 by .085.

Percents with fractions in them are the hardest to use. Sometimes you can change the percent to a simple fraction.

EXAMPLE: Find $83\frac{1}{3}$% of 108.

$83\frac{1}{3}\% = \frac{5}{6}$

Step 1.   Change $83\frac{1}{3}$% to a fraction.
(Look at the list on page 139.)

$\frac{5}{6} \times 108 =$

Step 2.   Multiply 108 by $\frac{5}{6}$.

$\frac{5}{\cancel{6}} \times \frac{\overset{18}{\cancel{108}}}{1} = \frac{90}{1} = 90$

You will not always know what fraction a percent is equal to. Divide the percent by 100 to get the fraction. Then multiply the fraction by the other number.

EXAMPLE: Find $28\frac{4}{7}$% of 56.

$\dfrac{28\frac{4}{7}}{100}$

Step 1.   Write $28\frac{4}{7}$ as the numerator and 100 as the denominator.

$28\frac{4}{7} \div 100 =$

Step 2.   Divide $28\frac{4}{7}$ by 100.

$\frac{200}{7} \div \frac{100}{1} =$

$\frac{200}{7} \times \frac{1}{100} = \frac{2}{7}$

Step 3.   Multiply 56 by $\frac{2}{7}$.

$\frac{2}{\cancel{7}} \times \frac{\overset{8}{\cancel{56}}}{1} = \frac{16}{1} = 16$

There is a shorter way to work this problem. Change the percent to an improper fraction. Then multiply the improper fraction by the other number in the numerator and by 100 in the denominator.

$28\frac{4}{7} = \frac{200}{7}$

Step 1.   Change $28\frac{4}{7}$ to an improper fraction.

Step 2.   Multiply $\frac{200}{7}$ by $\frac{56}{100}$.

$\frac{\overset{2}{\cancel{200}}}{\cancel{7}} \times \frac{\overset{8}{\cancel{56}}}{\cancel{100}} = \frac{16}{1} = 16$

Make sure that you understand both ways to solve the problem. Then you can decide which way is easier for you.

Solve each problem. Answers are on page 175.

1. 80% of 45 =   25% of 64 =   30% of 120 =

2. 50% of 82 =   35% of 60 =   64% of 125 =

3. 5% of 40 =   18% of 300 =   24% of 75 =

4. 300% of 52 =   150% of 34 =   500% of 17 =

5. 4.5% of 200 =   6.2% of 800 =   3.8% of 500 =

6. 50.2% of 250 =   37.5% of 120 =   9.8% of 600 =

7. 0.75% of 1,200 =   1.5% of 900 =   0.4% of 420 =

8. $62\frac{1}{2}$% of 56 = $16\frac{1}{3}$% of 120 = $6\frac{1}{4}$% of 160 =

9. $87\frac{1}{2}$% of 96 = $22\frac{2}{9}$% of 180 = $33\frac{1}{3}$% of 156 =

10. $\frac{3}{4}$% of 800 = $42\frac{6}{7}$% of 140 = $\frac{2}{5}$% of 1,500 =

Finding Percents Applications

These problems give you a chance to apply your skills in finding a percent of a number. When you find a percent of a number, you find a part of that number. The answer will have the same label as the number in the problem. For example, if you find a percent of $35, the answer will be measured in dollars. **Answers are on page 175.**

1. The town of Midvale spends 50% of its budget on education. Last year their budget for the year was $3,500,000. How much did they spend on education?

2. Rita took a math test with 80 problems. She got 75% of the problems right. How many problems did she get right?

3. George and Margaret want to buy a house for $36,500. They have to make a down payment of 15%. How much is the down payment for the house?

4. The Mayflower Theater holds 300 people. Friday 85% of the seats were filled. How many people were at the theater on Friday?

5. Colette owes $160 on her credit card. In a month she has to pay a fee of 1.5% of the amount she owes. Find the amount of the fee on the $180.

6. Richard had 60 tickets to sell to his club's dance. In two days he sold $83\frac{1}{3}$% of the tickets. How many tickets did he sell during those two days?

7. The sales tax in Steve's state is 8%. Steve bought a shirt for $14.50. How much was the sales tax on the shirt?

In many percent problems, you will use two operations to find an answer. First find a percent of a number. Then add or subtract this new amount with the old amount in the problem.

EXAMPLE: Dave bought a coat on sale. The coat used to cost $65. He bought it for 20% off the old price. How much did Dave pay for the coat?

$20\% = \dfrac{20}{100} = \dfrac{1}{5}$ *Step 1.*   Change 20% to a fraction.

$\dfrac{1}{\cancel{5}_1} \times \dfrac{\cancel{65}^{13}}{1} = \$13$ *Step 2.*   Multiply $65 by $\frac{1}{5}$.

$\$65 - \$13 = \$52$ *Step 3.*   Subtract $13 from $65.

Read each problem carefully to decide if you need to add or subtract.

8. Selma makes $245 a week. Her employer takes out 15% of her pay for taxes and Social Security. How much does Selma take home each week?

9. Frank took a math test with 40 problems. He got 85% of the problems right. How many problems did he get wrong?

10. Don bought records for $15.60. The sales tax in his state is 5%. How much did the records cost including sales tax?

11. In June Petra's phone bill was $24.50. In July her bill was 30% more because of long distance calls. How much was her July phone bill?

12. On a normal work day about 24,000 people ride the buses in Midvale. Monday was a holiday, and the number of riders was down 35%. How many people rode the buses on Monday?

Interest is money someone pays for using someone else's money. A bank pays you interest for using your money in a savings account. You pay a bank interest for using the bank's money on a loan.

To find interest, multiply the principal by the rate by the time.

The **principal** is the money you borrow or save.

The **rate** is the percent of the interest.

The **time** is the number of years.

EXAMPLE: Find the interest on $900 at 7% annual interest for one year.

$7\% = \dfrac{7}{100}$ *Step 1.* Change 7% to a fraction.

$\dfrac{\overset{9}{\cancel{900}}}{1} \times \dfrac{7}{\cancel{100}} \times 1 = \$63$ *Step 2.* Multiply the principal by the rate by the time.

Find the interest for each of the following. Answers are on page 175.

13. $800 at 6% annual interest for one year. $600 at 3.5% annual interest for one year.

14. $400 at 12% annual interest for one year. $450 at 4.8% annual interest for one year.

15. $1,500 at $5\frac{1}{2}$% annual interest for one year. $720 at 9% annual interest for one year.

16. $5,000 at $8\frac{3}{4}$% annual interest for one year. $4,800 at 6.5% annual interest for one year.

17. $1,200 at 1.5% annual interest for one year. $3,600 at 15% annual interest for one year.

When the time period for interest is not one year, change the time to a fraction of a year.

EXAMPLE: Find the interest on $900 at 7% annual interest for 8 months.

$7\% = \dfrac{7}{100}$

$\dfrac{8}{12} = \dfrac{2}{3}$

$\dfrac{\overset{3}{\cancel{\underset{1}{900}}}}{1} \times \dfrac{7}{\underset{1}{\cancel{100}}} \times \dfrac{2}{\cancel{3}} = \$42$

Step 1. Change 7% to a fraction.

Step 2. Change 8 months to a fraction. Write 8 in the numerator and 12 months (one whole year) in the denominator.

Step 3. Multiply the principal by the rate by the time.

When the time period is more than one year, write the time as a mixed number. For example, one year and six months $= 1\frac{6}{12} = 1\frac{1}{2}$.

Find the interest for each of the following.

18. $500 at 4% annual interest for 6 months.

$840 at 8% annual interest for 4 months.

19. $2,000 at 6% annual interest for 9 months.

$960 at $4\frac{1}{2}$% annual interest for 3 months.

20. $480 at 5% annual interest for 5 months.

$2,400 at 10% annual interest for 1 year and 8 months.

21. $360 at 3.5% annual interest for 1 year and 4 months.

$3,600 at 9% annual interest for 1 year and 9 months.

22. $4,000 at $2\frac{3}{4}$% annual interest for 2 years.

$5,000 at 8.5% annual interest for 2 years and 6 months.

Finding What Percent One Number Is of Another

When you studied fractions, you learned how to find what part one number is of another. You made a fraction with the part as the numerator and the whole as the denominator. The steps are almost the same for finding what percent one number is of another. Make a fraction with the part as the numerator and the whole as the denominator. Then change the fraction to a percent.

EXAMPLE: 36 is what % of 48?

$\frac{36}{48} = \frac{3}{4}$

Step 1.   Make a fraction with the part (36) over the whole (48).

$\frac{3}{4} \times 100\% =$

Step 2.   Reduce the fraction by 12.

$\frac{3}{\cancel{4}_1} \times \frac{\cancel{100}^{25}}{1} = \frac{75}{1} = 75\%$

Step 3.   Change $\frac{3}{4}$ to a percent.

Solve each problem. Answers are on page 175.

1. 28 is what % of 56?   32 is what % of 128?   16 is what % of 80?

2. 50 is what % of 75?   27 is what % of 72?   44 is what % of 55?

3. 6 is what % of 42?   30 is what % of 75?   24 is what % of 80?

4. 35 is what % of 45?   40 is what % of 48?   140 is what % of 160?

5. 14 is what % of 20?   120 is what % of 160?   24 is what % of 72?

6. 21 is what % of 126?   150 is what % of 240?   45 is what % of 300?

More Finding Percents Applications

These problems give you a chance to apply your skills in finding what percent one number is of another. In these problems compare one number to another by making a fraction. Then change the fraction to a percent. The answers to these problems are all measured in percents. **Answers are on page 175.**

1. Mark and Heather Green had to drive 540 miles to visit their children. Heather drove 162 miles. What percent of the distance did Heather drive?

2. Mike borrowed $1,600 to buy a used car. He had to pay $240 interest on the loan. The interest was what percent of the loan?

3. The Soto family spends $950 a month. They spend an average of $76 a month for gas, electricity, and telephone. These expenses make up what percent of the Sotos' budget?

4. Last year Patty made $232 a week. This year she got a $29 a week raise. Her raise was what percent of her old weekly salary?

5. There are 240 workers at the Midvale Utility Plant. 200 of the workers are men. What percent of the workers at the plant are men?

6. The city of Newport spent about $16 million last year for police and fire protection. It spent about $50 million on education. The amount it spent for police and fire protection was what percent of the amount it spent on education?

In some problems you have to compare the difference between two amounts to an original amount. First subtract to find the difference. Then make a fraction with the difference as the numerator and the original amount as the denominator. Change the fraction to a percent.

EXAMPLE: In 1979 Jane paid $180 a month for rent. In 1980 she paid $207 a month. By what percent did Jane's rent increase?

$$\begin{array}{r} \$207 \\ -\ \ 180 \\ \hline \$\ \ 27 \end{array}$$

Step 1. Find how much Jane's rent went up. Subtract the old rent from the new rent.

$$\frac{27}{180} = \frac{3}{20}$$

Step 2. Make a fraction with the difference ($27) over the original rent ($180) and reduce.

$$\frac{3}{\cancel{20}} \times \frac{\overset{5}{\cancel{100}}}{1} = 15\%$$

Step 3. Change $\frac{3}{20}$ to a percent.

EXAMPLE: Joe bought a jacket on sale for $52. Before the sale the jacket cost $65. Find the percent of discount on the original price.

$$\begin{array}{r} \$65 \\ -\ \ 52 \\ \hline \$13 \end{array}$$

Step 1. Find the difference (the discount). Subtract the new price from the old price.

$$\frac{13}{65} = \frac{1}{5}$$

Step 2. Make a fraction with the difference ($13) over the original price ($65).

$$\frac{1}{\cancel{5}} \times \frac{\overset{20}{\cancel{100}}}{1} = 20\%$$

Step 3. Change $\frac{1}{5}$ to a percent.

Be sure you understand these examples before you try the next problems. Notice that the original amount in the first example was the bigger number. The original in the second example was the smaller number.

7. Last year Nancy made $4.80 an hour. This year she makes $5.28 an hour. By what percent did her wage increase?

8. Mr. Walek runs a shoe store. He pays an average of $15 for a pair of shoes. He charges his customers an average of $21. By what percent does he mark up the price of a pair of shoes?

9. Two years ago Larry bought gas for $.84 a gallon. This year he pays $1.26 a gallon. By what percent did the price go up?

10. Jeff bought a T.V. on sale for $187. Before the sale the T.V. cost $220. Find the percent of discount on the original price.

11. Last year there were 16 students in Mr. Green's night school math class. This year there are 22 students. By what percent did the number of students increase?

12. At noon on April 21 the temperature was 72°. At 9 P.M. the temperature was 48°. By what percent did the temperature drop?

13. When Carlos started working five years ago, he made $8,400 a year. Now he makes $14,700 a year. By what percent did Carlos' salary go up?

14. Before Bill went on a diet he weighed 220 pounds. Now he weighs 176 pounds. What percent of his weight did he lose?

15. Two years ago an average of 24 people went to the meetings of the Lakeview Tenants' Association. Now about 48 people go to the meetings. By what percent did the number of people at the meetings increase?

Finding a Number When a Percent of It Is Given

When you found a percent of a number, you first changed the percent to a fraction or a decimal. Then you multiplied by the fraction or decimal. You found **a part of** the number.

Some problems are just the opposite. In these problems you already have the part. You have to find the whole that the part came from. You work these problems the opposite way. Change the percent to a fraction or a decimal. Then **divide** the number you have by the fraction or decimal.

EXAMPLE: 40% of what number is 18?

If we had the missing number, 40% multiplied by the number would give us 18. To find the missing number, divide 18 by 40%. (When you go on to study algebra, you will often use this method to solve problems. It is called using opposite operations.)

Using a fraction:

$40\% = \dfrac{40}{100} = \dfrac{2}{5}$ *Step 1.* Change 40% to a fraction.

$18 \div \dfrac{2}{5} =$ *Step 2.* Divide 18 by $\frac{2}{5}$.

$\dfrac{\overset{9}{\cancel{18}}}{1} \times \dfrac{5}{\cancel{2}} = \dfrac{45}{1} = 45$

Using a decimal:

$40\% = 40\% = .4$ *Step 1.* Change 40% to a decimal.

$\begin{array}{r} 4\,5. \\ .4\overline{)21.0} \end{array}$ *Step 2.* Divide 18 by .4 .

To check this problem, find 40% of 45. The answer should be 18.

$40\% = \dfrac{2}{5} \qquad \dfrac{2}{5} \times 45 = \dfrac{2}{\cancel{5}} \times \dfrac{\overset{9}{\cancel{45}}}{1} = \dfrac{18}{1} = 18$

Solve and check each problem. Answers are on page 176.

1. 60% of what number is 27? $33\frac{1}{3}$% of what number is 36?

2. 75% of what number is 48? $83\frac{1}{3}$% of what number is 35?

3. 60% of what number is 96? $12\frac{1}{2}$% of what number is 14?

4. 25% of what number is 19? 80% of what number is 52?

5. $16\frac{2}{3}$% of what number is 21? 50% of what number is 126?

6. 4.5% of what number is 54? 35% of what number is 84?

7. 37.5% of what number is 90? 9.2% of what number is 69?

8. 20% of what number is 18? 88% of what number is 132?

9. 3.4% of what number is 51? $66\frac{2}{3}$% of what number is 102?

More Percent Applications

These problems give you a chance to apply your skills in finding a number when a percent of the number is given. The number you find will be measured in the same units as the number in the problem. Give each answer the correct label. **Answers are on page 176.**

1. Tony had to pay $337.50 interest on a loan. The interest rate on his loan was 13.5%. What was the total amount of the loan?

2. Judy got 117 problems right on a math test in her night school class. She got a score of 78% right. How many problems were on the test?

3. 65% of the workers at the Midvale Utility Company voted to strike. 156 workers voted to strike. How many workers are there in the company?

4. Don sells record players on commission. One week he made $289.80 in commissions. He gets a 9% commission on everything he sells. Find the total value of the record players Don sold that week.

5. In 1970, 378 people belonged to the Lakeview Tenants' Association. The number of people who belonged to the group in 1970 was only 45% of the number who belonged in 1980. How many people were in the group in 1980?

6. Mr. and Mrs. Cooke pay $4,035 a year on their mortgage. That amount is 30% of their income. How much do the Cooke's make in a year?

7. Ellen Johnson ran for school board president in her town. She got 527 votes. Her votes were 62% of the total number of votes. How many people voted in the election?

8. The number of color T.V.s in the U.S. in 1970 was 78% of the number in 1980. In 1970 there were 230.1 million color T.V.s. How many were there in 1980?

9. Mr. and Mrs. Miller spend an average of $60.80 a week on food for their family. Food is 38% of the Miller's budget. Find their total budget for the week.

10. The members of the Lakeview Tenants' Association started to sell tickets for their spring fair. In a week they sold 540 tickets. The tickets they sold that week are 40% of the number they hope to sell. How many tickets do they hope to sell?

11. The sales tax in Ira's state is 8%. Ira had to pay $3.52 tax on a portable radio. Find the price of the radio.

12. In 1970 there were about 16 million people living in the New York City area. Experts think this is about 80% of the number that will be living in the area in 1990. If the experts are right, how many people will live in the New York area in 1990?

13. In 1975 Mrs. Smith paid 54¢ for a loaf of bread. This is 75% of the amount she paid for a loaf in 1980. How much did Mrs. Smith pay for a loaf of bread in 1980?

Percent Review

These problems will help you find out if you need to review the percent section of this book. When you finish, look at the chart to see which pages you should review.

1. Change each decimal to a percent.

   .3 =   .09 =   .455 = $.08\frac{1}{4} =$

2. Change each percent to a decimal.

   48% =   3% = $7\frac{1}{2}\% =$ 265% =

3. Change each fraction to a percent.

   $\frac{9}{10} =$ $\frac{1}{16} =$ $\frac{5}{12} =$ $\frac{4}{5} =$

4. Change each percent to a fraction.

   85% =   2.8% = $22\frac{2}{9}\% =$ $26\frac{2}{3}\% =$

5. 16% of 125 =   6. 40% of 75 =

7. 370% of 90 =   8. 4.8% of 800 =

9. $4\frac{1}{6}\%$ of 960 =   10. $66\frac{2}{3}\%$ of 129 =

11. The Lopez family makes $14,800 a year. They spend 25% of their income on rent. How much rent do they pay in a year?

12. Sam bought a shirt for $9.80. He had to pay 5% sales tax. How much did Sam pay for the shirt including tax?

13. Find the interest on $1,600 at 5% annual interest for one year.

14. Find the interest on $720 at 15% annual interest for 9 months.

15. 45 is what % of 75?

16. 36 is what % of 54?

17. 48 is what % of 192?

18. 60 is what % of 96?

19. Last year Fred made $280 a week. This year he got a $14 a week raise. His raise is what percent of his old weekly salary?

20. Before Joe went on a diet he weighed 180 pounds. Now he weighs 153 pounds. What percent of his weight did he lose?

21. 30% of what number is 57?

22. 75% of what number is 108?

23. 48% of what number is 60?

24. $33\frac{1}{3}$% of what number is 24?

25. There were 240,000 people living in Central County in 1970. The 1970 population was 75% of the number living there in 1980. How many people lived in Central County in 1980?

26. Dan paid $2.94 in tax on a bicycle for his daughter. The tax in his state is 6%. What was the cost of the bicycle?

Check your answers on page 176. Then turn to the review pages for the problems you missed.

If you missed problems	review pages
1 to 4	135 to 139
5 to 10	140 to 142
11 to 14	143 to 146
15 to 20	147 to 150
21 to 26	151 to 154

Answers

pages 2-6
1. thousand
2. million   thousand
3. million   thousand
4. million   thousand
5. 802   **6.** 40,530   **7.** 3,608,900
8. 60,070,050
9. 898   **10.** 88,679   **11.** 957   **12.** 1,367
13. 146   **14.** 157   **15.** 6,018   **16.** 16,584
17. 3,521   **18.** 6,482
19. $27.66   **20.** 270 pounds
21. 34   **22.** 540   **23.** 91,011   **24.** 38
25. 567   **26.** 32,827   **27.** 226   **28.** 10,704
29. 554   **30.** 1,578   **31.** 27,093   **32.** 3,876
33. $3.35   **34.** 563 tickets
35. 146   **36.** 1,692   **37.** 1,743   **38.** 80,916
39. 672   **40.** 4,704   **41.** 4,720   **42.** 168,665
43. 2,520   **44.** 25,917   **45.** 499,992   **46.** 162,955
47. 870   **48.** 6,000   **49.** 48,000
50. 2,622 miles   **51.** $195
52. 85   **53.** 74   **54.** 93r5   **55.** 47r2
56. 786r3   **57.** 49   **58.** 296r20   **59.** 58
60. 673   **61.** 406r5   **62.** 56   **63.** 43r25
64. 18 gallons
65. $11

page 7
1. 400   **2.** 90
3. 7   **4.** 30,000
5. 1,000   **6.** 800,000
7. 6,000,000   **8.** 2,000
9. 500   **10.** 70,000,000

page 8
1. hundred
2. thousand
3. thousand
4. million   thousand
5. thousand
6. thousand   hundred
7. million   thousand   hundred
8. 308   **9.** 261,000
10. 90,024   **11.** 4,170,000
12. 804,500   **13.** 60,300
14. 11,207,000

page 9
1. 10   13   8   16   11   16   6   14   11   5
2. 8   9   8   11   4   17   9   10   11   12

3. 13 13 9 12 3 10 7 12 7 9
4. 11 6 8 2 4 9 15 13 5 8
5. 11 6 15 11 9 13 6 3 15 7
6. 11 7 4 8 14 12 4 14 9 11
7. 13 7 5 2 10 5 15 6 6 14
8. 7 1 9 10 5 18 12 3 12 10
9. 9 16 8 8 10 10 10 14 7 12

pages 10-11
1. 89 88 96 58 98 98 79
2. 99 66 87 89 78 87 99
3. 669 793 969 869 957 598
4. 779 695 387 869 789 485
5. 9,665 9,878 9,477 9,488 9,299
6. 19,797 95,367 58,798 78,595 89,789
7. 557 458 998
8. 668 697 799
9. 2,449 8,859 6,778
10. 5,698 6,579 7,984
11. 4,996 8,566 2,684
12. 1,876 5,087 3,367
13. 6,697 3,999 7,889

pages 12-15
1. 101 102 100 180 144 101 103
2. 103 130 142 112 133 145 130
3. 114 162 141 80 110 73 143
4. 158 171 134 223 150 122 106
5. 147 106 180 205 154 153 153
6. 136 195 206 134 154 134 136
7. 400 313 460 702 485 869
8. 533 470 1,031 638 350 412
9. 1,413 1,142 503 1,201 683 1,115
10. 1,093 1,470 847 477 1,846 1,676
11. 1,608 1,245 1,137 2,650 2,095 886
12. 2,551 4,316 4,290 7,613 9,113
13. 5,351 3,506 2,554 7,450 2,849
14. 16,490 11,478 9,514 7,141 12,588
15. 576 929 **16.** 1,388 412
17. 5,919 9,046 **18.** 9,103 1,962
19. 2,660 5,426 **20.** 5,807 7,383
21. 9,597 5,367 **22.** 1,079 1,135
23. 1,271 1,091 **24.** 3,204 4,573
25. 17,754 22,366 **26.** 34,288 96,189
27. 19,381 20,201 **28.** 82,009 31,540

pages 16-17
1. 802 people **2.** $202.50
3. $25.97 **4.** 711 seats
5. $293.42
6. 891 calories **7.** $87.45
8. 3,361 albums **9.** 44,504 people
10. 18,332 votes **11.** $69.61
12. 850 miles

page 18

1.	6	3	6	8	6	0	9	9	5
2.	0	4	2	8	7	1	1	2	1
3.	3	1	4	6	3	6	0	3	6
4.	2	1	6	2	6	8	0	4	4
5.	1	9	4	1	3	9	5	5	0
6.	8	4	7	2	9	0	5	8	7
7.	9	3	9	2	4	7	5	2	7
8.	8	8	6	7	3	7	8	4	0
9.	1	7	2	2	5	0	5	7	4
10.	8	9	8	3	5	3	5	0	9

page 19

1.	62	43	12	25	30	56	61
2.	441	133	505	541	103	230	
3.	641	404	222	232	521	220	
4.	11,021	71,113	31,144	73,110	90,502		
5.	15,221	53,141	23,371	23,541	12,141		

pages 20-21

1.	26	87	18	69	78	57	38
2.	75	34	59	33	19	69	28
3.	29	48	19	39	8	16	28
4.	27	26	9	69	36	17	18
5.	68	38	14	27	24	29	75
6.	479	358	587	499	149	296	
7.	568	154	268	102	186	293	
8.	5,679	6,856	4,649	5,869	2,896		
9.	2,788	2,969	2,786	1,343	1,229		
10.	2,421	743	3,287	5,245	2,189		
11.	6,037	29,099	76,559	21,158	78,473		
12.	9,459	18,966	16,821	14,857	28,691		

page 22

1.	544	877	239	159	349	577
2.	565	198	115	189	288	328
3.	347	239	29	358	29	257
4.	439	378	94	546	227	38

page 23

1.	374	446	311	286	192	24
2.	1,085	853	5,817	1,603	1,984	1,672
3.	2,744	662	3,589	50	3,924	3,553
4.	57,025	30,987	23,341	4,633	65,983	

page 24

1.	743	174	651
2.	126	486	253
3.	5,730	1,319	4,977
4.	8,663	5,207	4,298
5.	2,445	1,608	3,538
6.	13,177	70,289	62,239
7.	12,339	10,784	21,118
8.	71,785	46,522	19,619

pages 25-26
1. $197.15   **2.** $27,720   **3.** 177 people
4. 188 miles   **5.** 144 years   **6.** $376,740
7. $3.11   **8.** $1,125   **9.** $3,820
10. 333 people   **11.** $39.15   **12.** 71,877 people
13. 7,350,000 barrels   **14.** $445

pages 27-28
1.	21	12	77	6	48	0	54	2
2.	36	108	0	55	120	36	4	28
3.	16	42	80	18	33	32	7	55
4.	88	0	24	14	35	84	27	36
5.	72	81	48	0	88	8	10	63
6.	60	16	10	22	30	110	96	48
7.	20	15	60	9	9	4	18	44
8.	96	110	72	64	70	16	24	21
9.	121	40	30	32	0	45	40	36
10.	0	72	99	24	84	33	54	24
11.	60	28	4	90	132	20	0	18
12.	27	50	12	66	5	2	20	8
13.	48	40	80	44	25	50	12	42
14.	49	77	14	30	0	3	22	120
15.	10	0	12	1	6	99	70	72
16.	100	12	35	56	66	63	18	45
17.	36	56	0	108	60	90	144	15
18.	48	132	10	24	20	30	40	6

page 30
1.	328	189	148	810	208	568	168
2.	287	240	55	248	129	288	630
3.	2,488	2,408	1,296	648	7,299	923	
4.	308	1,808	1,460	1,688	4,888	2,880	
5.	2,466	32,088	63,777	6,468	18,099		

page 31
1.	1,323	3,036	516	3,400	2,788	2,263	1,376
2.	14,314	9,399	11,088	25,092	7,014	7,308	
3.	274,284	76,146	535,236	313,728	97,356	91,803	
4.	639,969	3,088,892	592,128	1,398,930	1,086,492		

pages 32-34
1.	336	455	522	312	291	132	114
2.	444	738	296	340	156	198	162
3.	405	114	672	301	552	380	385
4.	4,272	1,435	1,125	3,056	1,884	4,410	
5.	4,632	2,037	1,870	2,904	5,304	6,503	
6.	4,932	5,536	1,374	4,325	2,562	4,074	
7.	1,368	2,730	1,075	3,024	3,542	3,520	2,574
8.	2,736	5,456	986	1,908	1,276	1,530	2,812
9.	3,402	3,995	3,762	532	2,144	2,537	2,262
10.	1,440	3,750	5,520	1,360	7,280	1,660	3,780
11.	4,020	7,920	1,920	1,350	6,320	3,480	750
12.	6,020	5,670	2,920	3,400	6,560	1,520	3,180
13.	7,334	17,280	64,998	44,616	28,242	60,516	

14. 20,784   23,086   35,916   30,552   29,770   80,178
15. 49,014   31,464   7,756   61,992   59,616   21,707
16. 232,824   90,036   110,592   203,490   567,648   156,000
17. 144,210   51,168   224,048   242,946   172,431
18. 217,281   328,572   270,088   712,810   276,444
19. 935,830   691,566   2,916,620   996,086   2,140,140

page 35
1. 4,208   3,311   2,436
2. 6,597   2,922   1,945
3. 3,760   2,590   4,140
4. 2,175   3,312   3,854
5. 14,112   62,952   28,050
6. 77,970   317,756   139,272

page 36
1. 920   1,360   80   2,040
2. 870   9,070   4,600   380
3. 1,600   28,800   9,000   4,300
4. 700   6,800   49,300   6,000
5. 17,000   9,000   208,000   46,000
6. 123,000   14,000   785,000   3,000
7. 2,900   40,000   20,000   3,600

pages 37-38
1. $9,869.60   **2.** $6.76   **3.** 288 miles
4. 306 miles   **5.** 180 inches
6. $152.25   **7.** 288 pounds   **8.** $1,500
9. 86¢   **10.** $6.48   **11.** $358.80
12. 1,335 words   **13.** 232 miles
14. 384 ounces

page 39
1. 6   1   8   6   1   0   7
2. 3   9   4   2   6   5   7
3. 2   1   9   4   3   6   8
4. 0   9   3   1   9   9   6
5. 9   4   8   7   1   7   8
6. 3   7   7   2   4   6   8
7. 1   8   9   0   5   5   7
8. 6   4   4   7   8   9   6
9. 4   3   3   8   7   5   2
10. 1   9   6   5   2   1   4
11. 3   0   5   4   2   2   5
12. 2   8   5   2   0   3   5

page 41
1. 47   23   85   56   44
2. 77   36   45   69   87
3. 79   58   64   88   92
4. 276   839   366   851   928
5. 444   233   863   938   654

pages 42-43
1. 41r5   39r1   22r6   52r4   78r2

2.	60r3	56r1	44r5	73r4	66r2
3.	93r2	77r5	37r4	99r1	74r1
4.	67r2	81r2	50r3	47r8	65r3
5.	91r1	70r6	42r5	67r3	76r3
6.	55r1	81r6	49r2	80r5	37r3
7.	192r6	721r3	304r7	586r2	
8.	640r5	817r2	920r6	455r2	
9.	272r7	794r1	968r1	582r2	
10.	312r2	564r3	541r4	848r2	
11.	912r1	702r7	386r1	907r1	
12.	855r2	727r5	440r5	689r2	
13.	615r1	449r7	706r3	480r4	

pages 44-46

1.	7	4	8	6	
2.	8	7	6	5	
3.	9r30	5r15	4r12	8r15	
4.	48	56	65	52	
5.	91	77	65	89	
6.	38	46	37	59	
7.	73r10	81r25	36r8	26r40	
8.	29r60	50r16	82r5	73r15	
9.	582	346	506	853	
10.	691r30	704r13	377r18	426r20	
11.	8	6	9	4	
12.	7	6	9	8	
13.	32	38	62	53	
14.	47r200	55r150	36r120	60r425	
15.	88	47	63	52	

page 47

1.	234	418	609
2.	512r6	388r2	490r3
3.	91	75	37
4.	53r20	66r15	85r24
5.	96	48	77

pages 48-49

1. 89 boxes   2. $230   3. 12 hours
4. $24.50   5. 35 pounds
6. $1,850   7. 236 boxes   8. 487 miles
9. 23 miles   10. 11 inches   11. 7¢
12. 27 hours   13. $123.60

pages 50-54

1. thousand
2. million   thousand
3. million   thousand   hundred
4. million   thousand
5. 540   6. 15,206
7. 4,120,008   8. 90,076,800
9. 799   10. 49,588   11. 459
12. 9,987   13. 161   14. 216
15. 11,401   16. 49,932

17. 7,417 **18.** 36,738
19. $33.83 **20.** 148 employees
21. 53 **22.** 322 **23.** 62,132
24. 74 **25.** 366 **26.** 22,195
27. 229 **28.** 32,123 **29.** 727
30. 3,345 **31.** 85,218 **32.** 41,114
33. $16.50 **34.** 116 people
35. 168 **36.** 2,048 **37.** 2,108
38. 210,672 **39.** 432 **40.** 4,672
41. 5,810 **42.** 146,328 **43.** 5,640
44. 20,410 **45.** 202,950 **46.** 159,354
47. 1,230 **48.** 1,600 **49.** 9,000
50. $9.45 **51.** 89,760 feet
52. 73 **53.** 49 **54.** 86r5 **55.** 49r4
56. 480r6 **57.** 93 **58.** 627r10 **59.** 54
60. 495 **61.** 802r6 **62.** 74 **63.** 86r15
64. 16 hours **65.** 36 months

pages 55-58

1. $\frac{1}{6}$ $\frac{3}{4}$ $\frac{5}{9}$ $\frac{3}{5}$

2. $\frac{7}{12}$ **3.** $\frac{3}{20}$

4. $\frac{3}{4}$ $\frac{5}{12}$

5. $\frac{8}{3}$ $\frac{2}{2}$ $\frac{7}{3}$

6. $6\frac{2}{5}$

7. $\frac{1}{9}$ $\frac{3}{4}$ $\frac{3}{13}$ $\frac{1}{15}$

8. $\frac{21}{56}$ $\frac{18}{75}$ $\frac{20}{48}$ $\frac{70}{100}$

9. $2\frac{2}{5}$ $4\frac{1}{3}$ 1 $6\frac{2}{3}$

10. $\frac{17}{3}$ $\frac{24}{7}$ $\frac{22}{5}$ $\frac{41}{5}$

11. $\frac{5}{8}$ $\frac{3}{10}$ $\frac{5}{6}$ $\frac{2}{3}$

12. $\frac{1}{5}$ **13.** $\frac{5}{9}$

14. $\frac{3}{5}$ **15.** $\frac{3}{4}$ **16.** $6\frac{1}{4}$ **17.** $\frac{2}{9}$

18. $\frac{13}{14}$ **19.** $11\frac{1}{24}$ **20.** $7\frac{19}{20}$ **21.** $10\frac{19}{60}$

22. $67\frac{3}{8}$ inches **23.** $5\frac{11}{16}$ pounds

24. $\frac{1}{5}$ **25.** $\frac{1}{4}$ **26.** $\frac{7}{24}$ **27.** $3\frac{4}{9}$

28. $5\frac{2}{5}$ **29.** $2\frac{3}{8}$ **30.** $2\frac{11}{18}$ **31.** $3\frac{11}{15}$

32. $24\frac{1}{4}$ hours **33.** $21\frac{5}{8}$ pounds

34. $\frac{10}{27}$ **35.** $\frac{5}{21}$ **36.** $\frac{4}{5}$ **37.** $6\frac{2}{3}$

38. 70 **39.** $\frac{2}{3}$ **40.** $10\frac{1}{2}$ **41.** 6

42. $3.90 **43.** $7,200

44. $1\frac{1}{2}$ **45.** $9\frac{1}{3}$ **46.** $7\frac{1}{2}$ **47.** $\frac{1}{24}$

48. $\frac{5}{22}$ **49.** $1\frac{1}{3}$ **50.** $1\frac{1}{4}$ **51.** $1\frac{1}{7}$

52. $1.30 **53.** $20\frac{1}{4}$ inches

pages 59-60

1. $\frac{2}{3}$ $\frac{1}{4}$ $\frac{4}{6}$ $\frac{2}{4}$

2. $\frac{5}{8}$ $\frac{1}{2}$ $\frac{3}{5}$ $\frac{2}{5}$

3. $\frac{3}{10}$ $\frac{3}{8}$ $\frac{4}{6}$ $\frac{5}{9}$

4. $\frac{3}{4}$ 5. $\frac{7}{10}$ 6. $\frac{21}{60}$ 7. $\frac{127}{1000}$

8. $\frac{19}{36}$ 9. $\frac{2}{7}$

page 61

1. $\frac{6}{7}$ $\frac{4}{5}$ $\frac{8}{200}$ 2. $\frac{19}{5}$ $\frac{12}{9}$ $\frac{15}{15}$ $\frac{24}{4}$

3. $8\frac{4}{7}$ $2\frac{3}{20}$ $3\frac{8}{9}$ $10\frac{1}{2}$

4. $\frac{12}{2}$ $6\frac{3}{5}$ $\frac{20}{7}$ $2\frac{9}{16}$ $3\frac{5}{8}$ $\frac{41}{4}$

page 63

1. $\frac{1}{5}$ $\frac{1}{9}$ $\frac{1}{7}$ $\frac{1}{8}$ $\frac{1}{6}$

2. $\frac{6}{13}$ $\frac{5}{16}$ $\frac{1}{3}$ $\frac{7}{10}$ $\frac{8}{11}$

3. $\frac{4}{5}$ $\frac{1}{2}$ $\frac{1}{5}$ $\frac{1}{10}$ $\frac{1}{25}$

4. $\frac{5}{6}$ $\frac{4}{5}$ $\frac{5}{8}$ $\frac{2}{5}$ $\frac{2}{5}$

5. $\frac{3}{11}$ $\frac{11}{12}$ $\frac{3}{4}$ $\frac{5}{9}$ $\frac{2}{3}$

6. $\frac{4}{7}$ $\frac{3}{5}$ $\frac{5}{9}$ $\frac{1}{2}$ $\frac{1}{12}$

7. $\frac{1}{4}$ $\frac{1}{21}$ $\frac{7}{11}$ $\frac{7}{8}$ $\frac{7}{9}$

8. $\frac{3}{4}$ $\frac{1}{2}$ $\frac{4}{5}$ $\frac{4}{5}$ $\frac{3}{7}$

9. $\frac{6}{7}$ $\frac{4}{5}$ $\frac{3}{7}$ $\frac{1}{2}$ $\frac{9}{13}$

page 64

1. $\frac{18}{24}$ $\frac{8}{36}$ $\frac{56}{80}$ $\frac{14}{35}$ $\frac{25}{40}$

2. $\frac{54}{63}$ $\frac{3}{36}$ $\frac{30}{50}$ $\frac{24}{54}$ $\frac{16}{22}$

3. $\frac{24}{36}$ $\frac{16}{72}$ $\frac{42}{60}$ $\frac{7}{56}$ $\frac{36}{45}$

4. $\frac{24}{32}$ $\frac{3}{45}$ $\frac{30}{55}$ $\frac{40}{60}$ $\frac{36}{40}$

page 65

1. $2\frac{1}{2}$ $3\frac{1}{4}$ $5\frac{2}{3}$ $8\frac{3}{5}$ $4\frac{4}{9}$

2. $3\frac{7}{10}$ 6 $2\frac{4}{7}$ $5\frac{8}{9}$ 8

3. $1\frac{5}{6}$ $3\frac{5}{12}$ 5 $2\frac{3}{4}$ $3\frac{2}{7}$

4. $5\frac{1}{3}$ $6\frac{1}{2}$ $9\frac{5}{8}$ $2\frac{1}{2}$ $4\frac{5}{6}$

page 66

1. $\frac{20}{3}$ $\frac{7}{2}$ $\frac{23}{4}$ $\frac{27}{10}$ $\frac{7}{6}$

2. $\frac{29}{8}$ $\frac{29}{3}$ $\frac{30}{7}$ $\frac{41}{6}$ $\frac{19}{10}$

3. $\frac{25}{3}$ $\frac{29}{12}$ $\frac{37}{4}$ $\frac{39}{8}$ $\frac{27}{7}$

4. $\frac{15}{2}$ $\frac{19}{16}$ $\frac{38}{9}$ $\frac{43}{12}$ $\frac{24}{5}$

page 68

1. $\frac{3}{5}$ $\quad$ $\frac{5}{12}$ $\quad$ $\frac{8}{15}$ $\quad$ $\frac{7}{9}$

2. $\frac{21}{25}$ $\quad$ $\frac{4}{7}$ $\quad$ $\frac{1}{4}$ $\quad$ $\frac{2}{3}$

3. $\frac{2}{5}$ $\quad$ $\frac{5}{9}$ $\quad$ $\frac{3}{4}$ $\quad$ $\frac{3}{8}$

4. $\frac{5}{6}$ $\quad$ $\frac{2}{3}$ $\quad$ $\frac{5}{8}$ $\quad$ $\frac{7}{8}$

5. $\frac{3}{8}$ $\quad$ $\frac{5}{12}$ $\quad$ $\frac{3}{4}$

6. $\frac{29}{36}$ $\quad$ $\frac{9}{20}$ $\quad$ $\frac{2}{3}$

page 69

1. $\frac{1}{4}$ $\qquad$ 2. $\frac{4}{5}$ $\qquad$ 3. $\frac{1}{7}$

4. $\frac{2}{9}$ $\qquad$ 5. $\frac{1}{8}$ $\qquad$ 6. $\frac{3}{10}$

pages 70-71

1. $\frac{5}{7}$ $\quad$ $\frac{5}{6}$ $\quad$ $10\frac{7}{9}$ $\quad$ $10\frac{5}{8}$ $\quad$ $8\frac{9}{10}$

2. $\frac{13}{20}$ $\quad$ $\frac{3}{5}$ $\quad$ $11\frac{3}{4}$ $\quad$ $12\frac{9}{16}$ $\quad$ $14\frac{13}{24}$

3. $\frac{17}{18}$ $\quad$ $\frac{6}{11}$ $\quad$ $12\frac{7}{15}$ $\quad$ $12\frac{25}{32}$ $\quad$ $16\frac{23}{30}$

4. $\frac{1}{3}$ $\quad$ $\frac{1}{2}$ $\quad$ $7\frac{1}{2}$ $\quad$ $11\frac{1}{2}$ $\quad$ $10\frac{3}{5}$

5. $\frac{4}{5}$ $\quad$ $\frac{3}{4}$ $\quad$ $12\frac{5}{6}$ $\quad$ $13\frac{3}{4}$ $\quad$ $11\frac{7}{10}$

6. $\frac{1}{3}$ $\quad$ $\frac{4}{5}$ $\quad$ $13\frac{3}{4}$ $\quad$ $10\frac{3}{4}$ $\quad$ $8\frac{5}{8}$

7. $10\frac{1}{5}$ $\quad$ $13\frac{2}{3}$ $\quad$ $14\frac{1}{3}$ $\quad$ $11\frac{1}{3}$ $\quad$ 15

8. $9\frac{4}{11}$ $\quad$ $7\frac{2}{7}$ $\quad$ $15\frac{1}{2}$ $\quad$ $7\frac{1}{3}$ $\quad$ $16\frac{2}{3}$

9. $17\frac{2}{5}$ $\quad$ 11 $\quad$ $7\frac{1}{2}$ $\quad$ $11\frac{3}{10}$ $\quad$ $14\frac{1}{3}$

pages 72-75

1. $1\frac{1}{2}$ $\quad$ $1\frac{1}{9}$ $\quad$ $1\frac{5}{8}$ $\quad$ $1\frac{1}{10}$ $\quad$ $1\frac{1}{6}$

2. $1\frac{2}{15}$ $\quad$ $1\frac{1}{6}$ $\quad$ $1\frac{1}{16}$ $\quad$ $1\frac{3}{20}$ $\quad$ $1\frac{1}{6}$

3. $1\frac{3}{20}$ $\quad$ $1\frac{7}{24}$ $\quad$ $1\frac{13}{30}$ $\quad$ $1\frac{1}{14}$ $\quad$ $1\frac{5}{18}$

4. $\frac{29}{30}$ $\quad$ $1\frac{6}{35}$ $\quad$ $\frac{23}{30}$ $\quad$ $1\frac{9}{40}$ $\quad$ $\frac{35}{36}$

5. $1\frac{5}{18}$ $\quad$ $\frac{19}{24}$ $\quad$ $1\frac{7}{30}$ $\quad$ $1\frac{5}{36}$ $\quad$ $\frac{29}{48}$

6. $1\frac{11}{24}$ $\quad$ $1\frac{3}{20}$ $\quad$ $1\frac{1}{12}$ $\quad$ $\frac{43}{60}$ $\quad$ $1\frac{7}{40}$

7. $\frac{23}{24}$ $\quad$ $1\frac{11}{20}$ $\quad$ $1\frac{19}{40}$ $\quad$ $1\frac{11}{36}$ $\quad$ $1\frac{7}{9}$

8. $1\frac{13}{15}$ $\quad$ $1\frac{11}{28}$ $\quad$ $1\frac{11}{24}$ $\quad$ $1\frac{47}{60}$ $\quad$ $1\frac{13}{18}$

9. $15\frac{11}{24}$ $\quad$ $18\frac{17}{40}$ $\quad$ $17\frac{23}{28}$ $\quad$ $14\frac{7}{8}$

10. $9\frac{37}{60}$ $\quad$ $15\frac{7}{18}$ $\quad$ $16\frac{27}{40}$ $\quad$ $19\frac{31}{42}$

11. $17\frac{17}{36}$ $\quad$ $24\frac{7}{24}$

12. $21\frac{2}{15}$ $\quad$ $19\frac{13}{18}$

13. $25\frac{7}{16}$ $\quad$ $25\frac{17}{18}$

14. $20\frac{8}{15}$ $\quad$ $22\frac{21}{40}$

pages 76-77

1. $\frac{14}{15}$ of their income   2. $72\frac{5}{8}$ inches

3. $9\frac{7}{12}$ hours   4. $15\frac{3}{5}$ miles   5. $10\frac{7}{8}$ pounds

6. $\$3\frac{3}{4}$ million   7. $12\frac{1}{12}$ feet   8. $12\frac{5}{12}$ hours

9. $21\frac{13}{16}$ pounds   10. $2\frac{1}{4}$ hours   11. $9\frac{3}{16}$ pounds

page 78

1. $\frac{3}{5}$ $\frac{2}{5}$ $7\frac{1}{3}$ $5\frac{3}{7}$ $1\frac{2}{7}$

2. $\frac{3}{4}$ $\frac{1}{2}$ $9\frac{1}{2}$ $3\frac{3}{10}$ $2\frac{3}{5}$

3. $\frac{1}{3}$ $\frac{2}{3}$ $4\frac{1}{3}$ $3\frac{2}{5}$ $5\frac{1}{3}$

page 79

1. $\frac{5}{24}$ $\frac{3}{10}$ $7\frac{5}{12}$ $4\frac{2}{9}$ $4\frac{1}{3}$

2. $\frac{1}{10}$ $\frac{11}{21}$ $3\frac{3}{8}$ $2\frac{2}{15}$ $9\frac{19}{36}$

3. $\frac{7}{12}$ $\frac{5}{18}$ $1\frac{1}{4}$ $2\frac{5}{24}$ $2\frac{8}{15}$

pages 80-82

1. $3\frac{4}{7}$ $1\frac{1}{8}$ $3\frac{2}{9}$ $2\frac{1}{4}$ $4\frac{1}{2}$

2. $1\frac{3}{10}$ $6\frac{7}{12}$ $1\frac{7}{16}$ $1\frac{7}{15}$ $7\frac{9}{20}$

3. $4\frac{4}{7}$ $3\frac{2}{3}$ $4\frac{3}{5}$ $4\frac{1}{2}$ $1\frac{1}{2}$

4. $2\frac{3}{4}$ $7\frac{1}{3}$ $2\frac{2}{5}$ $8\frac{2}{3}$ $2\frac{1}{2}$

5. $2\frac{5}{6}$ $6\frac{17}{20}$ $3\frac{17}{24}$ $1\frac{5}{8}$ $3\frac{7}{18}$

6. $1\frac{5}{12}$ $1\frac{16}{21}$ $4\frac{2}{3}$ $6\frac{16}{28}$ $4\frac{15}{28}$

7. $4\frac{4}{5}$ $3\frac{27}{40}$ $2\frac{11}{30}$ $1\frac{8}{9}$ $6\frac{7}{9}$

8. $4\frac{17}{30}$ $3\frac{7}{20}$ $8\frac{1}{6}$ 9. $7\frac{13}{18}$ $2\frac{3}{4}$ $1\frac{17}{30}$

10. $3\frac{13}{21}$ $5\frac{3}{4}$ $9\frac{1}{2}$ 11. $6\frac{11}{40}$ $2\frac{29}{100}$ $4\frac{3}{5}$

pages 83-84

1. $\$2\frac{1}{8}$ million   2. $2\frac{3}{4}$ inches

3. $57\frac{7}{8}$ inches   4. $1\frac{3}{4}$ pounds

5. $4\frac{2}{5}$ miles

6. $\frac{1}{2}$ million people   7. $188\frac{1}{2}$ pounds

8. $\$1\frac{2}{5}$ million   9. $22\frac{3}{4}$ hours

10. $\frac{3}{20}$ of their income   11. $\frac{1}{6}$ hour

page 85

1. $\frac{2}{15}$ $\frac{9}{40}$ $\frac{12}{35}$ $\frac{4}{15}$

2. $\frac{5}{48}$ $\frac{21}{40}$ $\frac{8}{45}$ $\frac{1}{56}$

3. $\frac{35}{72}$ $\frac{9}{28}$ $\frac{8}{27}$ $\frac{5}{24}$

4. $\frac{8}{25}$ $\frac{16}{75}$ $\frac{30}{77}$ $\frac{25}{72}$

page 87

1. $\frac{8}{21}$ $\frac{5}{14}$ $\frac{7}{12}$ $\frac{2}{5}$
2. $\frac{1}{12}$ $\frac{1}{6}$ $\frac{4}{15}$ $\frac{5}{26}$
3. $\frac{2}{27}$ $\frac{4}{5}$ $\frac{12}{35}$ $\frac{3}{28}$
4. $\frac{10}{63}$ $\frac{1}{9}$ $\frac{2}{5}$ $\frac{9}{28}$
5. $\frac{6}{25}$ $\frac{1}{8}$ $\frac{3}{5}$ $\frac{1}{20}$
6. $\frac{3}{4}$ $\frac{3}{20}$ $\frac{4}{9}$ $\frac{3}{4}$
7. $\frac{3}{5}$ $\frac{4}{15}$ $\frac{3}{11}$ $\frac{1}{18}$
8. $\frac{5}{12}$ $\frac{1}{3}$ $\frac{5}{32}$ $\frac{3}{5}$

page 88

1.   6   6 $6\frac{2}{3}$ $1\frac{1}{2}$
2.   8 $8\frac{1}{3}$ 4 $1\frac{1}{2}$
3. $1\frac{3}{4}$ $2\frac{1}{2}$ 6 $9\frac{3}{5}$
4. $7\frac{1}{2}$ $17\frac{1}{2}$ $\frac{2}{3}$ $3\frac{1}{2}$

pages 89-90

1. $3\frac{1}{3}$ $1\frac{7}{8}$ $6\frac{1}{4}$ $1\frac{1}{6}$
2.   28   21 $6\frac{1}{2}$ $19\frac{1}{2}$
3. $3\frac{1}{5}$ 2 $1\frac{2}{3}$ $2\frac{2}{3}$
4.   18 $13\frac{1}{2}$ 15 $21\frac{1}{2}$
5.   4 $1\frac{1}{5}$ 5 $2\frac{2}{5}$
6. $1\frac{3}{7}$ $\frac{5}{6}$ $\frac{13}{14}$ $2\frac{1}{10}$
7. $1\frac{13}{20}$ $13\frac{1}{2}$ $9\frac{4}{5}$ $6\frac{3}{4}$
8. $1\frac{3}{5}$ 4 $6\frac{3}{4}$ $2\frac{1}{3}$
9.   9 $5\frac{5}{6}$ $1\frac{1}{2}$ 10
10.   5 $10\frac{1}{2}$ 32   3

pages 91-92

1. $47   **2.** 76 ounces   **3.** $25.20
4. $1\frac{7}{8}$ pounds   **5.** 77¢   **6.** 392 people
7. $2,500   **8.** 48 cars   **9.** $37.70
10. $29\frac{1}{3}$ feet   **11.** $5,200   **12.** $35.70

pages 93-95

1. $1\frac{1}{5}$ $1\frac{1}{3}$ $1\frac{3}{7}$ $7\frac{1}{2}$
2. $\frac{2}{3}$ $\frac{3}{5}$ $\frac{2}{3}$ $\frac{3}{4}$
3. $\frac{3}{5}$ $1\frac{7}{8}$ $\frac{7}{10}$ $3\frac{3}{4}$
4.   16   10 $10\frac{1}{2}$ $7\frac{1}{2}$

5.   8 $3\frac{1}{3}$ 8   10

6.   12 $26\frac{2}{3}$ $16\frac{1}{2}$ $18\frac{2}{3}$

7. $9\frac{1}{3}$ 12 $13\frac{1}{3}$ $3\frac{1}{2}$

8. $2\frac{2}{5}$ $5\frac{1}{3}$ 5 $3\frac{1}{3}$

9.   6 $4\frac{1}{2}$ 10 $2\frac{4}{5}$

page 96

1. $\frac{1}{10}$ $\frac{2}{45}$ $\frac{1}{16}$ $\frac{1}{12}$

2. $\frac{1}{21}$ $\frac{3}{50}$ $\frac{5}{24}$ $\frac{4}{45}$

3. $\frac{3}{10}$ $\frac{1}{2}$ $\frac{1}{6}$ $1\frac{1}{5}$

4. $\frac{3}{14}$ $\frac{3}{4}$ $\frac{2}{3}$ $\frac{7}{12}$

pages 97-98

1. $\frac{1}{10}$ $\frac{3}{16}$ $\frac{1}{6}$ $\frac{1}{6}$

2. $\frac{5}{24}$ $\frac{3}{8}$ $\frac{3}{7}$ $\frac{12}{35}$

3. $\frac{5}{39}$ $\frac{1}{4}$ $\frac{3}{8}$ $\frac{2}{15}$

4. $2\frac{1}{2}$ $\frac{2}{3}$ $2\frac{2}{5}$ $4\frac{2}{3}$

5.   4 $3\frac{1}{2}$ $2\frac{1}{4}$ $1\frac{2}{3}$

6.   6 $1\frac{1}{7}$ $1\frac{3}{4}$ $6\frac{1}{4}$

7. $1\frac{1}{2}$ $\frac{5}{6}$ $2\frac{2}{3}$ $1\frac{1}{2}$

8. $\frac{2}{3}$ $4\frac{1}{2}$ $2\frac{1}{2}$ $\frac{4}{5}$

9. $\frac{5}{12}$ $1\frac{1}{3}$ $2\frac{1}{7}$ $2\frac{2}{5}$

pages 99-100

1. $1.80   **2.** $8\frac{1}{8}$ inches   **3.** 22 cans

4. 9 pieces   **5.** $5\frac{1}{4}$ pounds   **6.** $4.50

7. $6\frac{1}{4}$ minutes   **8.** 48 lots

9. 6 bookcases   **10.** 12 pans   **11.** 14 rows

pages 101-104

1. $\frac{1}{4}$ $\frac{5}{6}$ $\frac{1}{3}$ $\frac{2}{5}$

2. $\frac{13}{1000}$ **3.** $\frac{7}{24}$ **4.** $\frac{4}{11}$ $\frac{3}{8}$

5. $\frac{9}{2}$ $\frac{6}{6}$ $\frac{13}{3}$ **6.** $8\frac{4}{7}$

7. $\frac{1}{12}$ $\frac{5}{9}$ $\frac{1}{3}$ $\frac{2}{25}$

8. $\frac{27}{33}$ $\frac{32}{56}$ $\frac{48}{54}$ $\frac{10}{32}$

9. $6\frac{2}{3}$ 1 $2\frac{2}{5}$ $3\frac{1}{3}$

10. $\frac{44}{7}$ $\frac{23}{4}$ $\frac{29}{6}$ $\frac{28}{3}$

11. $\frac{7}{10}$ $\frac{1}{3}$ $\frac{1}{6}$ $\frac{2}{3}$

12. $\frac{3}{8}$ **13.** $\frac{4}{5}$

14. $\frac{5}{7}$ **15.** $\frac{4}{5}$ **16.** $11\frac{1}{3}$ **17.** $1\frac{1}{3}$

18. $1\frac{1}{18}$ **19.** $19\frac{11}{36}$ **20.** $19\frac{1}{20}$ **21.** $13\frac{13}{24}$

22. $8\frac{11}{16}$ pounds   **23.** $9\frac{4}{5}$ miles

24. $\frac{3}{8}$ **25.** $\frac{1}{5}$ **26.** $\frac{13}{30}$ **27.** $4\frac{7}{12}$

28. $1\frac{3}{4}$ **29.** $5\frac{11}{15}$ **30.** $4\frac{2}{3}$ **31.** $3\frac{1}{2}$

32. $\$\frac{17}{20}$ million   **33.** $20\frac{3}{8}$ inches

34. $\frac{15}{56}$ **35.** $\frac{5}{24}$ **36.** $\frac{1}{3}$ **37.** $5\frac{3}{5}$

38. 22   **39.** $\frac{9}{10}$ **40.** 8   **41.** 6

42. $4,300   **43.** 260 employees

44. $\frac{3}{4}$ **45.** 12   **46.** $4\frac{1}{2}$ **47.** $\frac{7}{24}$

48. $\frac{7}{20}$ **49.** $1\frac{1}{2}$ **50.** $\frac{1}{2}$ **51.** $\frac{5}{18}$

52. $3.30   **53.** 24 bags

pages 105-108

1. **a.** five tenths,   six and three tenths
   b. seven hundredths,   nine and twelve hundredths
   c. thirty-five thousandths,   twelve and eight thousandths
   d. twenty-nine ten-thousandths,   three and sixteen millionths

2. **a.** .9   **b.** .022   **c.** .0016
   d. .00086   **e.** 5.02
   f. 60.093   **g.** 407.000033

3. **a.** .9   .052   .304
   b. .3   .52   .081

4. **a.** .4   .707   .22
   b. .82   .05   .502

5. **a.** $\frac{1}{20}$ $\frac{9}{10}$ $3\frac{4}{5}$ $6\frac{3}{8}$
   b. $\frac{6}{125}$ $6\frac{13}{200}$ $\frac{1}{250}$ $10\frac{1}{5}$

6. **a.** .8 $.58\frac{1}{3}$ .55 $.28\frac{4}{7}$
   b. .9   .48 $.11\frac{1}{9}$ .18

7. 1.7   **8.** 2.217   **9.** .6286

10. 37.93   **11.** 15.92   **12.** 14.318

13. 203.2 million   **14.** 12.5 hours

15. .12   **16.** 2.65   **17.** 7.862

18. 11.392   **19.** 6.126   **20.** 13.18

21. 8.95 kilograms   **22.** .3 billion barrels

23. 2.1   **24.** 12.04   **25.** 1.014

26. 64.5   **27.** .2356   **28.** .9591

29. $6.72   **30.** $202.50

31. 2.9   **32.** .37   **33.** 5.8

34. 23   **35.** 50   **36.** 40

37. $2.90   **38.** 5 miles

page 110
1. three tenths,   four and two tenths
2. six hundredths,   eight and seven hundredths
3. fifteen thousandths,   ten and three thousandths
4. sixteen ten-thousandths,   sixty and seven ten-thousandths
5. three hundred four hundred-thousandths,   three and nine millionths
6. .3   5.04
7. .013   30.7
8. .02   90.06
9. .0012   .000016
10. 8.000304

page 111
1. .95   .3   .07
2. .4   .04   .061
3. .64   .33   .564
4. .2   .72   .1013
5. .7   .43   .201
6. .302   .505   .77
7. .82   .79   .3303
8. .6306   .9   .2

page 112
1. $\frac{2}{25}$ $\frac{2}{5}$ $6\frac{6}{25}$ $9\frac{3}{8}$
2. $\frac{5}{8}$ $\frac{1}{1000}$ $3\frac{1}{400}$ $7\frac{21}{50}$
3. $\frac{11}{2000}$ $\frac{3}{4}$ $8\frac{7}{20}$ $4\frac{22}{25}$
4. $\frac{12}{25}$ $\frac{29}{40}$ $3\frac{9}{25}$ $11\frac{1}{200}$
5. $\frac{7}{200}$ $\frac{3}{400}$ $6\frac{8}{125}$ $2\frac{1}{25,000}$

page 113
1. .25   .6   .7 $.66\frac{2}{3}$ $.83\frac{1}{3}$
2. .45   .15 $.37\frac{1}{2}$ or .375   .64 $.22\frac{2}{9}$
3. .3 $.41\frac{2}{3}$ $.57\frac{1}{7}$ .5   .16
4. $.33\frac{1}{3}$ .4 $.62\frac{1}{2}$ or .625 $.56\frac{1}{4}$ $.08\frac{1}{3}$

pages 114-115
1. 1.289   1.899
2. 1.7   1.29
3. 2.077   .9917
4. 1.183   1.1686
5. 1.0219   .7957
6. .767   .581
7. 72.07   10.106
8. 76.746   6.667
9. 6.358   35.3
10. 12.527   44.443
11. 78.07   8.875
12. 7.3945   10.7682
13. 260.872   28.049

page 116
1. 56.1°   2. $3.05 million   3. 16 miles
4. 4.5 billion   5. 37,174.1 miles
6. 103.5°   7. 12.75 hours

page 117
1. 5.641   11.65   .218
2. .282   3.941   .016
3. .853   5.401   .224
4. 16.11   3.925   5.76
5. .018   29.2   .251
6. .027   1.5694   14.2
7. 7.79   .221   3.596

page 118
1. .5 million square miles
2. 198.7 million   3. 4,353.4 miles
4. .009   5. .25 meter
6. .8 million   7. .15 meter
8. 10.5 pounds

page 120
1. 18.2   7.38   .24
2. .072   5.46   .22
3. 4.938   .6818   33.48
4. 18.86   16.34   3.168
5. .0188   .00042   .00144
6. 17.4   1,391.2   56.32
7. 10.4   49.167   2.0102
8. 38.7   24.48   6,111
9. .03168   40.85   .0254

page 121
1. 81 kilograms   2. $64.50
3. 179.8 miles   4. $16.74
5. $3.22   6. 162.56 centimeters
7. 9 miles   8. 20¢   9. $14.45

page 122
1. 3.7   .52   .038
2. 3.29   .624   .67
3. .027   2.8   .039
4. 40.3   .071   .62

page 124
1. 3.4   .56   7.8
2. .47   9.2   57
3. .37   6.3   .58
4. .54   1.8   93
5. 470   26   59
6. 66   7.7   .64
7. .09   33   60
8. 327   19.6   3.85
9. .91   .53   50

page 125
1. 15   40   650
2. 12   250   500
3. 550   40   50
4. 4   4,200   60

pages 126-127
1. $2.70   2. .63 meter
3. 70 kilograms
4. 25 gallons   5. $.43
6. $4.65   7. 21.6 miles
8. .346 kilogram   9. 9.6 pounds
10. 55 miles   11. .49 meter

pages 128-131
1. a. six tenths,   eight and one tenth
   b. thirteen hundredths,   two and five hundredths
   c. four thousandths,   nine and one hundred eight thousandths
   d. six ten-thousandths,   seven and fifty-two ten-thousandths
2. a. .4   b. .018   c. .0005   d. .000013
   e. 1.08   f. 85.056   g. 220.0006
3. a. .5   .2   .9
   b. .52   .6   .25
4. a. .87   .11   .79
   b. .041   .28   .43
5. a. $\frac{3}{10}$ $3\frac{3}{4}$ $4\frac{4}{25}$ $\frac{1}{400}$
   b. $\frac{3}{2000}$ $2\frac{3}{50}$ $\frac{5}{8}$ $9\frac{8}{25}$
6. a. $.91\frac{2}{3}$ .2 $.12\frac{1}{2}$ or .125   .16
   b. .06 $.88\frac{8}{9}$ $.06\frac{1}{4}$ $.14\frac{2}{7}$
7. 1.5   8. 1.716   9. .9172
10. 27.96   11. 15.95   12. 20.429
13. 7.25 feet   14. 10.7 million
15. .51   16. 3.67   17. 6.073
18. 10.491   19. 6.248   20. 2.274
21. .85 meter   22. 7.3 years
23. 4.23   24. 6.46   25. 1.634
26. 130.2   27. .1332   28. 2.1385
29. $41.65   30. 19.2 kilometers
31. 4.8   32. .37   33. 6.9
34. .074   35. 200   36. 30
37. $1.60   38. 3.6 pounds

pages 132-134
1. 60%   6%   24.8% $3\frac{1}{3}$%
2. .5   .07 $.05\frac{1}{4}$ 3.25

3. 30% $12\frac{1}{2}$% or 12.5% $42\frac{6}{7}$%   16%

4. $\frac{9}{25}$ $\frac{17}{200}$ $\frac{5}{16}$ $\frac{5}{8}$

5. 21   **6.** 54   **7.** 108   **8.** 21

9. 17   **10.** 60

11. 51 questions   **12.** $28.90

13. $144   **14.** $69

15. 80%   **16.** $33\frac{1}{3}$%   **17.** 75%   **18.** $83\frac{1}{3}$%

19. 20%   **20.** 12%   **21.** 60   **22.** 260

23. 140   **24.** 150

25. 25 students   **26.** $185

page 136

1. 65%   6%   4.5% $16\frac{2}{3}$%

2. 80%   25% $6\frac{1}{4}$%   82%

3. .4%   20%   3%   36%

4. 280%   50%   62.5%   400%

5. .55   .08   .125   .02

6. .064 $.33\frac{1}{3}$ .6   2.25

7. .9   .004   .2   5

8. $.62\frac{1}{2}$ 3 $.01\frac{2}{3}$ .114

pages 137-138

1. 70%   60% $11\frac{1}{9}$%   18%

2. 32% $37\frac{1}{2}$% or 37.5% $33\frac{1}{3}$%   50%

3. $16\frac{2}{3}$% $28\frac{4}{7}$%   19% $62\frac{1}{2}$% or 62.5%

4. $41\frac{2}{3}$%   45% $6\frac{1}{4}$%   25%

5. $\frac{7}{20}$ $\frac{1}{50}$ $\frac{6}{25}$ $\frac{3}{10}$

6. $\frac{11}{25}$ $\frac{3}{50}$ $1\frac{1}{2}$ $\frac{3}{100}$

7. $\frac{4}{25}$ $4\frac{1}{2}$ $\frac{3}{20}$ $\frac{24}{25}$

8. $\frac{6}{125}$ $\frac{21}{200}$ $\frac{1}{2500}$ $\frac{11}{400}$

9. $\frac{1}{8}$ $\frac{5}{6}$ $\frac{3}{7}$ $\frac{1}{12}$

10. $\frac{5}{9}$ $\frac{7}{8}$ $\frac{1}{16}$ $\frac{1}{6}$

page 139

50% = .5 = $\frac{1}{2}$ 20% = .2 = $\frac{1}{5}$

40% = .4 = $\frac{2}{5}$

25% = .25 = $\frac{1}{4}$ 60% = .6 = $\frac{3}{5}$

75% = .75 = $\frac{3}{4}$ 80% = .8 = $\frac{4}{5}$

$12\frac{1}{2}$% = $.12\frac{1}{2}$ or .125 = $\frac{1}{8}$ 10% = .1 = $\frac{1}{10}$

$37\frac{1}{2}\% = .37\frac{1}{2} \text{ or } .375 = \frac{3}{8}$ $30\% = .3 = \frac{3}{10}$

$62\frac{1}{2}\% = .62\frac{1}{2} \text{ or } .625 = \frac{5}{8}$ $70\% = .7 = \frac{7}{10}$

$87\frac{1}{2}\% = .87\frac{1}{2} \text{ or } .875 = \frac{7}{8}$ $90\% = .9 = \frac{9}{10}$

$33\frac{1}{3}\% = .33\frac{1}{3} = \frac{1}{3}$ $16\frac{2}{3}\% = .16\frac{2}{3} = \frac{1}{6}$

$66\frac{2}{3}\% = .66\frac{2}{3} = \frac{2}{3}$ $83\frac{1}{3}\% = .83\frac{1}{3} = \frac{5}{6}$

page 142
1. 36   16   36
2. 41   21   80
3. 2   54   18
4. 156   51   85
5. 9   49.6   19
6. 125.5   45   58.8
7. 9   13.5   1.68
8. 35   20   10
9. 84   40   52
10. 6   60   6

pages 143-146
1. $1,750,000   2. 60 problems
3. $5,475   4. 255 people
5. $2.70   6. 50 tickets
7. $1.16
8. $208.25   9. 6 problems
10. $16.38   11. $31.85
12. 15,600 people
13. $48 $21
14. $48 $21.60
15. $82.50 $64.80
16. $437.50 $312
17. $18 $540
18. $10 $22.40
19. $90 $10.80
20. $10 $400
21. $16.80 $567
22. $220 $1062.50

page 147
1. 50%   25%   20%
2. $66\frac{2}{3}\%$ $37\frac{1}{2}\%$ 80%
3. $14\frac{2}{7}\%$ 40%   30%
4. $77\frac{7}{9}\%$ $83\frac{1}{3}\%$ $87\frac{1}{2}\%$
5. 70%   75% $33\frac{1}{3}\%$
6. $16\frac{2}{3}\%$ $62\frac{1}{2}\%$ 15%

pages 148-150
1. 30%   2. 15%   3. 8%
4. $12\frac{1}{2}\%$ 5. $83\frac{1}{3}\%$ 6. 32%

   7. 10%   **8.** 40%   **9.** 50%

  10. 15%   **11.** $37\frac{1}{2}$%   **12.** $33\frac{1}{3}$%

  13. 75%   **14.** 20%   **15.** 100%

pages 151-152
  1. 45   108
  2. 64   42
  3. 160   112
  4. 76   65
  5. 126   252
  6. 1,200   240
  7. 240   750
  8. 90   150
  9. 1,500   153

pages 153-154
  1. $2,500   **2.** 150 problems
  3. 240 members   **4.** $3,220
  5. 840 people   **6.** $13,450
  7. 850 people   **8.** 295 million
  9. $160   **10.** 1,350 tickets
  11. $44   **12.** 20 million
  13. 72¢

pages 155-157
  1. 30%   9%   45.5% $8\frac{1}{4}$%

  2. .48   .03   .07$\frac{1}{2}$ 2.65

  3. 90% $6\frac{1}{4}$% $41\frac{2}{3}$%   80%

  4. $\frac{17}{20}$ $\frac{7}{250}$ $\frac{2}{9}$ $\frac{4}{15}$

  5. 20   **6.** 30   **7.** 333

  8. 38.4   **9.** 40   **10.** 86

  11. $3,700   **12.** $10.29

  13. $80   **14.** $81

  15. 60%   **16.** $66\frac{2}{3}$%

  17. 25%   **18.** $62\frac{1}{2}$%

  19. 5%   **20.** 15%

  21. 190   **22.** 144

  23. 125   **24.** 72

  25. 320,000 people   **26.** $49